America Goes to War

Managing the Force During Times of Stress and Uncertainty

Bernard D. Rostker

Prepared for the Office of the Secretary of Defense

Approved for public release; distribution unlimited

NATIONAL DEFENSE RESEARCH INSTITUTE

The research described in this report was prepared for the Office of the Secretary of Defense (OSD). The research was conducted in the RAND National Defense Research Institute, a federally funded research and development center sponsored by the OSD, the Joint Staff, the Unified Combatant Commands, the Department of the Navy, the Marine Corps, the defense agencies, and the defense Intelligence Community under Contract W74V8H-06-2-0002.

Library of Congress Cataloging-in-Publication Data

Rostker, Bernard.
America goes to war : managing the force during times of stress and uncertainty / Bernard Rostker.
p. cm.
Includes bibliographical references.
ISBN 978-0-8330-3980-4 (pbk. : alk. paper)
1. Manpower—United States. 2. Military service, Voluntary—United States. 3. Draft—United States—History. 4. United States—Armed Forces—Recruiting, enlistment, etc. 5. Families of military personnel—Services for—United States. I. Title.

UA17.5.U5R67 2007
355.2'230973—dc22

2007009507

Published 2007 by the RAND Corporation
1776 Main Street, P.O. Box 2138, Santa Monica, CA 90407-2138
1200 South Hayes Street, Arlington, VA 22202-5050
4570 Fifth Avenue, Suite 600, Pittsburgh, PA 15213-2665
RAND URL: http://www.rand.org/
To order RAND documents or to obtain additional information, contact
Distribution Services: Telephone: (310) 451-7002;
Fax: (310) 451-6915; Email: order@rand.org

Preface

This report is the product of Phase I of a project to help the Department of Defense (DoD) manage its personnel during the current period of high levels of deployment and what has commonly become known as "stress." The Deputy Under Secretary of Defense (Program Integration) in the Office of the Under Secretary of Defense (Personnel and Readiness) sponsored this project.

In 1970, in the middle of the undeclared wars in Southeast Asia and with America engaged in combat in South Vietnam, Congress agreed to President Nixon's proposal to transition to an all-volunteer force. Since then, all branches of the military have relied on volunteers to meet their manpower needs. Currently with extended deployments in Iraq and Afghanistan, the military, particularly the reserve forces, are having some difficulties in recruiting. Questions have been raised concerning the viability of the all-volunteer force and how DoD can manage personnel during these times of stress and uncertainty. This report addresses these concerns, with particular attention to the history of conscription and volunteerism. The current policy to use financial incentives is well established in American history, as are efforts to restructure the military to meet the current threat. One thing unique to the all-volunteer force is the high proportion of military members who are married and have children. This presents new challenges, and a great many programs have been developed to help members and their families in these difficult times. Understanding which programs work is a particular challenge that is also addressed in this report.

This report should be of interest to anyone concerned with managing the force during periods of conflict and under conditions of stress and uncertainty. By providing a historical account of what has been tried before, the report will help build a firm base for current and future policies.

This research was sponsored by the Office of the Under Secretary of Defense (Personnel and Readiness) and conducted within the Forces and Resources Policy Center of the RAND National Defense Research Institute, a federally funded research and development center sponsored by the Office of the Secretary of Defense, the Joint Staff, the Unified Combatant Commands, the Department of the Navy, the Marine Corps, the defense agencies, and the defense Intelligence Community. Comments are welcome and may be addressed to Bernard Rostker. He can be reached by email at bernard_rostker@rand.org; by phone at 703-413-1100, extension 5481; or by mail at RAND Corporation, 1200 South Hayes Street, Arlington, VA 22202-5050.

For more information on RAND's Forces and Resources Policy Center, contact the Director, James Hosek. He can be reached by email at james_hosek@rand.org; by phone at 310-393-0411, extension 7183; or by mail at the RAND Corporation, 1776 Main Street, P.O. Box 2138, Santa Monica, CA 90407-2138. More information about RAND is available at www.rand.org.

Contents

Figures and Tables

Figures

Tables

Summary

Introduction

Currently, with extended deployments in Iraq and Afghanistan, the Army is having difficulties recruiting new non-prior-service personnel. Questions have been raised concerning the viability of the all-volunteer force and how the Department of Defense (DoD) can manage personnel during these times of stress and uncertainty. This report addresses these concerns, with particular attention to the history of conscription and volunteerism. It examines the history of the draft to try to understand when and under what conditions conscription has been used effectively to raise the manpower needed by the Army during wartime. The report also examines what other means besides conscription the Army could use to meet manpower demands. Specifically, what actions could be taken to increase the supply of volunteers or reduce the demand for new personnel? The report also looks at the many programs that have been developed to help military members and their families cope during difficult times, as well as the particular challenges of understanding which programs work.

To Draft or Not to Draft, That Is the Question

If there is to be a public debate over conscription, then it should consider under what conditions conscription has been used effectively to raise the manpower needed by the Army during wartime. The histories of Britain and France are most often used to spotlight the differences between countries that have favored volunteerism and those that have favored conscription and to help illustrate the conditions when conscription has been accepted.

Great Britain's Tradition

Great Britain, buttressed by the isolation afforded it by being separated from most of its adversaries by sea, was able to provide for the defense of the nation as it limited the power of the state in favor of a military force made up of volunteers. In 1916, the enormous manpower demands finally compelled Britain to enact a national conscription, but by March 1920, with occupational duty behind it, Britain ended its draft. It was not until the eve of World War II, April 27, 1939, that Britain again enacted national conscription. Between 1946 and 1960, Britain fought six colonial wars. By 1946, it was clear that the manpower needs of the armed forces were such that conscription could not end, and the wartime draft law was extended. In April 1957, the British government announced its decision to end conscription. By 1963, there were no conscripts serving in the British Army.

France's Tradition

The fundamental difference between Great Britain and France reflects the difference in philosophy of English philosopher John Locke and French philosopher Jean-Jacques Rousseau, the former emphasizing the rights and liberties of individual citizens and the latter a citizen's responsibilities to the state. In 1789, with foreign powers poised to restore the monarchy, the National Assembly reported, "Every citizen must be a soldier and every soldier a citizen, or we shall never have a constitution." Article 12 of The Declaration of the Rights of Men and Citizens of 1789 stipulates that "[t]he security of the rights of man and of the citizen requires public military forces," and Article 13 states that "common contribution is essential for the maintenance of the public forces." This was the prevailing doctrine for most of the next two hundred years. After 1989, with the end of the Cold War, and for the first time since 1871, no direct threat to its national territory, France started to move to an all-volunteer force. The two principles of "obligation and universality" on which the draft had been built were now coming into conflict with another principle—equality. France had a structural surplus of people eligible for national service beyond the needs of the military. As a result, fewer and fewer people actually served in the armed services. National service could be accomplished by serving for as little as ten months in the military or enrolling in one of five forms of civil service—or one could even claim to be a conscientious objector. The final move to an all-volunteer force came with the election of President Jacques Chirac in 1995 and from pressure from a reform movement that wanted a fully professional military.

Equity and the Prussian Model of Universal Selective Service

The original French model of conscription, with its emphasis on the obligation of all citizens to defend the revolution coexisting with provisions to allow a citizen to buy his way out of service, proved to be a clear contradiction; this model was finally corrected after the humiliating French defeat at the hands of Prussia in 1870. The modern model of universal military service developed by Prussia during and after the Napoleonic period proved so effective in allowing a country to mobilize its manpower and field a much larger army than might have been maintained as a standing force that by the end of the 19th century it was in wide use throughout the non-English-speaking world. It was the Prussian system of short-term conscripts backed by years of compulsory service in the reserves that defeated Denmark in 1864, Austria in 1866, and France in 1870. Thereafter, Austria (in 1868), France (in 1872), Italy (in 1873), Russia (in 1874), and Japan (in 1883) adopted, to one degree or another, the Prussian system of universal military training and selective service.

The American Tradition

The noted historian of the modern American draft, George Q. Flynn, suggests that the American tradition is rooted in its colonial past, when military service was seen

> less as a part of citizenship and more as a burden imposed by government. Operating under a heritage that stressed minimal

> government interference with individual choice, these cultures were able to sell military service only as a matter of national defense in an emergency. (Flynn, 2002, p. 3)

According to the Militia Act of 1792, "each and every free able-bodied white male . . . [between] the age of eighteen years and under the age of forty-five years . . . [was] enrolled in the militia" (O'Sullivan and Meckler, 1974, p. 36); however, by the time of the Mexican War (1846–1848), service in the militia had ceased to be compulsory.

Ironically, the first American Congress to pass a "national" conscription law was the Congress of the Confederate States of America. The North followed when, on March 3, 1863, President Lincoln signed the Union's first draft law, the Enrollment Act. The draft, however, was a despised institution because there was little sense of equal sacrifice. Following in the French tradition, wealthy men were able to buy their way out of service—commutation—or hire a substitute to serve in their stead. The draft riots in Boston, New York, and other Northern cities attested to its unpopularity. In the most perverse way, the draft was effective in the North, not because it brought in large numbers of people, but because it persuaded "elected officials to raise much higher bounties to entice men to enlist and thus avert the need for governmental coercion" (Chambers, 1987, p. 64).

Between the Civil War and World War I, including the Indian Wars and the Spanish-American War, America relied on volunteerism and the *new* volunteer militia of the states—the National Guard—to provide the manpower needed to defend the country. On April 2, 1917, President Wilson asked Congress for a Declaration of War. Four days later, the day Congress actually declared war on Germany, the president asked for a draft, and on May 18, 1917, he signed the Selective Service Act of 1917 into law. Unlike the Civil War draft, the new draft was widely accepted. Frederick Morse Cutler described the "marvelously complete response . . . the popular support and approval accorded the selective service," and how, on the day young men reported for registration, "a feeling of solemnity possessed all hearts; a holiday was declared; at the stated hour, church bells rang as though summoning men to worship" (Cutler, 1923, p. 174). While the law did not allow for bounties or personal substitution, it did provide for deferments based on essential work. The term Selective Service was used to capture the idea that, while all men of a specific age group—eventually 18 to 45 years of age—might be required to register, only some would be selected for military service in line with the total needs of the nation. The 72 percent of the armed forces that were draftees made a better case for equality of sacrifice than did those drafted during the Civil War. When the need for the mass army ended, however, so did the need for and legitimacy of the draft.

With war raging in Europe, conscription returned on September 16, 1940, when President Roosevelt signed the Selective Training and Service Act of 1940, the first peacetime conscription law in the history of the United States; the draft was sold as a democratic mechanism. In the preamble of the act, Congress declared that service should be shared according to a fair and just system.

Even before the end of World War II, however, with victory clearly ahead, Congress, under considerable pressure from the public, pressed President Truman to end the draft; the draft ended on March 31, 1947. In less than a year, however, the world situation had so deteriorated and the Army's experience with this version of an all-volunteer force had been so disastrous—with a requirement of 30,000 recruits a month, only 12,000 volunteers were coming forward—that President Truman asked for a resumption of the draft. By February 1949, however, inductions were suspended, and by the summer of 1949, the Associated Press reported that "unless an unforeseen emergency develops, the peacetime draft of manpower for the armed forces is expected to expire June 25, 1950" (Associated Press, 1949). On June 24, 1950, North Korean forces invaded South Korea. Three days later, Congress voted to extend military conscription.

The Korean War, and the war in Vietnam a decade later, did not mobilize and unite the country as the two World Wars had done, or at least had initially done; opinion polls showed that World War II was "unquestionably much more highly supported by the public than the Korean and Vietnam wars" (Mueller, 1973, p. 63). When the Korean armistice was signed, American troops remained in Korea and the draft stayed in place. In truth, this was not the end of a war but the end of a battle. The Cold War and draft continued. The problem of equity was captured in the title of one of numerous government studies of the period, *In Pursuit of Equity: Who Serves When Not All Serve?* (Marshall, 1967). In addition, LTG Lewis Hershey, the Director of Selective Service, would admit that "equity was unattainable" and that "we defer people . . . because we can't use them all" (Flynn, 1985, p. 218). The noted military sociologist James Burk found that

> the perception of inequities eroded public confidence in the draft. In 1966, for the first time since the question was asked, less than a majority (only 43 percent) believed that the draft was handled fairly in their community. Although the public still supported the draft, the problems protesters exposed raised serious questions about its operation during the Vietnam War. (Burk, 2001)

Burk's observations on inequities and public confidence echoed those of Alexis de Tocqueville more than a century before when he wrote, "The government may do almost whatever it pleases, provided it appeals to the whole community at once; it is the unequal distribution of the weight, not the weight itself, that commonly occasions resistance" (de Tocqueville, 1835, Chapter 23).

On October 17, 1968, at the height of the Vietnam War, the Republican candidate for president, Richard Nixon, addressed the nation on the subject of conscription. He called for an end to the draft because "a system of compulsory service that arbitrarily selects some and not others simply cannot be squared with our whole concept of liberty, justice and equality under the law" (Nixon, 1968). One week after taking office, Nixon told his Secretary of Defense, Melvin Laird, to "begin immediately to plan a special commission to develop a detailed plan of action for ending the draft" (Nixon, 1969). On February 21, 1970, the Commission on an All-Volunteer Armed Force (known as the Gates Commission) forwarded to President Nixon its recommendation to end conscription. The commission unanimously found the cost of an all-volunteer force was "a necessary price of defending our peace and security . . . [and that conscription] was intolerable when there is an alternative consistent with our basic national values" (Gates, 1970, p. 10). On September 28, 1971, President Nixon signed Public Law 92-129 and ushered in the era of the all-volunteer force.

When the Congress debated the end of conscription in 1970, the fate of the draft was very much uncertain. The issue made strange bedfellows. Some liberals in Congress, such as Senator Edward Kennedy (D-Mass.), feared that an all-volunteer force would be made up of the poor, black, and uneducated. Some conservatives, such as Senator John Stennis (D-Miss.), remembering the Army's experience in 1947, thought that a volunteer force would not attract sufficient numbers of recruits. Both sides agreed, however, that pay should be fair; as pay rose, so did the number of young men who volunteered. The end of the draft was certain when it became clear at market wages that there would be enough volunteers to man the force.

What History Tells Us

Since the time of the Civil War, the United States has used conscription four times. The draft was successful in meeting the manpower needs of the country twice, and twice volunteerism effectively replaced it. Conscription was successful during the two World Wars when the conflict had general popular support, the entire male population of military age was included (registered), and selection was judged to be fair and sacrifice perceived to be equal—equal in terms of the chance to serve, not in terms of the economic consequences of serving, or as the preamble of the 1940 draft law put it, "shared generally in accordance with a fair and just system of selective compulsory military training and service." When the cause did not enjoy the full support of the people, as in Vietnam, or the selection appeared to be random or biased with inequitable service, as in both the Civil War and the Vietnam War, conscription was unsuccessful.

American history suggests that conscription works only when (1) the cause enjoys overwhelming support among the general population and (2) there is a generally held belief that all are participating with equal sacrifice. Without both of these conditions in place, conscription has not been a viable way to raise the manpower needed by the military. Are the conditions right now for a return to conscription?

In the fall of 2004, an overwhelming majority of population—85 percent—replied "no" to the question, "Do you think the United States should return to a military draft at this time, or not?" (Gallup Brain, 2006b). Thus, it would appear that the current conflict does not enjoy the popular support needed to bring back the draft. Moreover, even if the military is not able to retain sufficient numbers of people to meet all its future requirements, it is unlikely that the numbers of men who would need to be drafted would be so large as to meet the criterion of "equal sacrifice" for the draft to be judged equitable.

To Go "Soldiering": Managing the Force Without a Draft

How can a volunteer force be maintained, even during periods of conflict? The government can (1) increase the supply of volunteers to either enlist or reenlist into the armed forces, (2) reduce the demand for manpower by restructuring the current force, or (3) try to ameliorate the most negative aspects of deployment and family separation that result in military personnel and their families making the decision to leave the military.

Increasing the Supply of Volunteers

While some may deride it, history has shown that volunteers increasingly respond to bonuses and pay, with higher levels of compensation resulting in a greater number of volunteers. The uses of "bounties," or what today are called bonuses, to encourage soldiers to both enlist and reenlist is as old as the Army itself. On January 19, 1776, General George Washington wrote to the Continental Congress urging its members to "give a bounty of six dollars and two thirds of a dollar to every able bodied effective man, properly clothed for the service, and having a good fire lock, with a bayonet" (as quoted in Assistant Secretary of Defense, Manpower and Reserve Affairs (ASD[M&RA]), 1967a, p. I.1). This first enlistment bonus eventually grew to $200 by the end of the war (Kreidberg and Henry, 1955, p. 14). Within weeks, on February 9, 1776, Washington, faced with the prospect of needing troops for another year, also noted that the Congress "would save money and have infinitely better troops if they were, even at the bounty of twenty, thirty or more dollars, to engage the men already enlisted" (ASD[M&RA], 1967a, p. I.2).

The notion that an all-volunteer force might be sustained during periods of conflict through the use of incentives was new and untried before the current war in Iraq. Crawford Greenewalt, a member of the Gates Commission, wrote to Thomas Gates in 1969 (as the commission was completing its work), "While there is a reasonable possibility that a peacetime armed force could be entirely voluntary, I am certain that an armed force involved in a major conflict could *not* be voluntary" (Greenewalt, 1969, emphasis in the original). Today, new financial incentives have been developed for both recruiting and retaining the personnel needed. The $420 billion National Defense Authorization Act of 2005 continued a full range of recruiting and retention bonuses, as well as extended health benefits for some reservists, and provided a new educational assistance program for the reserves tied to the Montgomery GI Bill. However, although using financial incentives to attract and retain military personnel seems to have been generally successful in allowing DoD to maintain the size of the active military, it comes at a substantial cost. David S.C. Chu, Under Secretary of Defense for Personnel and Readiness, remarked at a recent conference examining the rising cost of military personnel that

> it's important to keep in mind that the military compensation system, whatever its idiosyncrasies, does work reasonably well in producing the results that we want. . . . It's critical to keep in mind the compensation system is not an end of itself. . . . The system is, after all, an instrument to reach the results we want, which is to supply young Americans who are willing to take on some of the most difficult and demanding tasks that society might ask them to do. It's not the only reason they serve, but it's an important element of their decision to serve, and it's certainly important in their family's decision to support such service. . . . Cost is important and we want to be efficient, but it is critical to start with what . . . [we want] to achieve. (Horowitz and Bandeh-Ahmadi, 2004)

Reducing Demand by Transforming the Force

In 2004, Secretary of Defense Donald H. Rumsfeld told the Chairman of the House Armed Services Committee that the force was "stressed" because it was "not properly aligned or organized for the post–Cold War era" (Rumsfeld, 2004c). His solution was to (1) increase the size of the Army by 30,000 troops; (2) increase the number of deployable brigades from 33 to 43, with the goal of reducing the frequency of, and increasing the predictability of, deployments; and (3) "rebalance" skills between the active and reserve components.

Family Program to Ameliorate the Most Negative Aspects of Deployment

There is more to managing the force than just compensating people for their service or organizing the force to make sure that it can best meet current demands. Providing support services for service members and their families helps ameliorate the most negative aspects of deployments. However, traditionally, military life has not been "family friendly." Until World War II, with the exception of the period of World War I, the adage "If the Army had wanted you to have a wife, they would have issued you one" aptly summed up the service's attitude toward families. While the Cold War–era Army in no way resembled the pre–World War II organization of the same name—the postwar Army was many times the size of the prewar Army and had worldwide responsibilities—the Army's approach to addressing family concerns remained reactive and piecemeal. It took the move to the all-volunteer force to really change things.

On the eve of the all-volunteer force, the *Fiscal Year 1971 Department of the Army Historical Summary* made no mention of military families per se; it was only implied by concern that "the Army needs a total of 353,440 housing units for eligible families [when] available family housing on and off post total[s] 220,600 units" (Bell, 1973, p. 55). By 1978, however, the Army understood that its approach to its Quality of Life program, originally established to "improve services and activities for enlisted personnel in their daily life," needed to be expanded "to bolster community of life support activities" (Boldan, 1982, p. 91). Citing the all-volunteer force, the Army noted before the end of the draft that

> less than half of the soldiers were married. By the end of 1977, over 60 percent fell into that category, many more were sole parents, and a considerable number were married to other soldiers. The changing composition of the Army necessitated increased attention to community services to sustain morale and retain highly qualified personnel. (Boldan, 1982, p. 91)

In October 1980, the first Army Family Symposium was held, in Washington, D.C. On August 15, 1983, Army Chief of Staff John A. Wickham signed the *Army Family White Paper—The Army Family*. It provided for the annual *Army Family Action Plan*, the Army theme for 1984 ("Year of the Family"), and the establishment of installation-based Family Centers.

In 1990, service members were deployed overseas in Operations Desert Shield and Desert Storm, while their family members stayed behind. Despite the establishment of Army Community Service and 24-hour Family Assistance Centers at the seven stateside posts from which large numbers of troops deployed, and other programs, after-action reports showed that "[f]amily members of deployed service members had innumerable problems and questions, felt confused and abandoned, and often did not know where to turn to obtain resolution and answers" (Reeves, 1998). The Army established family support groups for every deployment and declared that "[q]uality of life is the Army's third highest priority, immediately behind readiness and modernization" (Reeves, 1998).

After the Gulf War and throughout the 1990s, ever-increasing deployments placed new demands on soldiers and their families; the Chief of Staff of the Army told Congress that Army families must be prepared to deal with the stress and uncertainty that deployment brings (West and Reimer, 1997).

In April 2002, DoD published *The New Social Compact* as a reciprocal understanding between the department and service members and their families. The document declared: "Service members and families together must dedicate themselves to the military lifestyle, while the American people, the President, and the Department of Defense must provide a supportive quality of life for those who serve" (Molino, 2002, p. 1). The compact provided an "overview of services' delivery systems and strategies" (Molino, 2002, pp. 103–113). Each of the services has since developed programs to mitigate family stress. With so many programs, however, it is hard to know which ones work and which ones do not work, and under what circumstances.

Effectiveness of Military Family Support Programs. From the very beginning of the modern family program, policymakers have been asking for some level of proof that family support programs are "cost-effective." The *Department of the Army Historical Summary* for FY 1981 noted, "The Quality of Life Program, after three years of planning and programming, at last received enough funds to make a noticeable difference for soldiers and their families" (Hardyman, 1988, p. 108). With costs projected to run $1.6 million over the next six years, the *Summary* commented: "Quality of life efforts have been handicapped in the competition for limited resources by the Army's inability to quantify the benefits derived from implementing the initiatives. There was no obvious way to measure soldiers' satisfaction and its effect on soldier commitment" (Hardyman, 1988, pp. 108–109). In 2004, policymakers were still looking for some way to determine which programs were cost-effective. The *First Quadrennial Quality of Life Review* reported that, despite the general recognition that quality of life "impacts the retention of service members and the readiness of the armed forces, . . . research that can inform policy on these issues is surprisingly inadequate" (DoD, 2004, p. 187).

Today, surveys and focus groups are the primary means we have for learning about these programs, but they provide an incomplete picture. Academic research that focuses on how people make the decision to stay or leave also provides little insight into where DoD should spend its money. Problems persist in determining the correct sampling design and the analytic and statistical approaches to follow. Overdue is a valid and reliable research design for the collection and analysis of information to assess the performance of the variety of family support programs.

Summary and Conclusion

Headlines notwithstanding, the all-volunteer force has done extremely well during these stressful and uncertain times. Commissioner Greenewalt's certainty in 1970 that "[a]n armed force involved in a major conflict could *not* be voluntary" (Greenewalt, 1969, emphasis in the original) has been proven wrong. History suggests that the conditions favorable to conscription—overwhelming support for the cause and equality of sacrifice—are not present today. The senior leaders in the administration and many in Congress are of an age at which former Secretary of Defense Casper Weinberger's words in 1987—"We know what the draft did to the social fabric of this country in the '60s" (as quoted in Chambers, 1987, p. 259)—are fair warning. The American military has been very resilient in finding ways to make the all-volunteer force work. However, a number of new and expanded compensation programs have been put in place, and retention has remained high; each of the services has restructured to provide additional personnel to meet the demands of new missions; and family programs have been expanded to mitigate stress.

As it has been from the beginning, the all-volunteer force remains fragile. Accordingly, DoD has provided a wide range of support programs to help service members and their families cope with the stress and uncertainty of heightened military operations and deployments. To date, increases in the operational tempo for active and reserve forces, including multiple tours in the combat areas of Afghanistan and Iraq, have not resulted in significant recruitment shortages or low retention. However, only time will tell.

Abbreviations

ACS	Army Community Services
AEF	Aerospace Expeditionary Force
AER	Army Emergency Relief
AFAP	Army Famiy Action Plan
AIP	Assignment Incentive Pay
ASD(M&RA)	Assistant Secretary of Defense, Manpower and Reserve Affairs
BAH	Base Allowance for Housing
BAS	Base Allowance for Subsistence
CBO	Congressional Budget Office
CDS	child development services
CHAMPUS	Civilian Health and Medical Program of the Uniformed Services
CNO	Chief of Naval Operations
CONUS	continental United States
CSG	carrier strike group
CY	calendar year
DMDC	Defense Manpower Data Center

DoD	Department of Defense
FAC	Family Assistance Center
FSA	Family Separation Allowance
FY	fiscal year
GAO	General Accounting Office; now, Government Accountability Office
HDP-L	Hardship Duty Pay—Location
IDP	Imminent Danger Pay
IRR	Individual Ready Reserve
JANSSC	Joint Army-Navy Selective Service Committee
JOPES	Joint Operations Planning and Execution System
MCCA	Military Child Care Act
MFRI	Military Family Research Institute
NCO	noncommissioned officer
OCONUS	outside the continental United States
OOTW	operation other than war
OPTEMPO	operational tempo
PCS	permanent change of station
PDC	permanent duty change
PERSTEMPO	personnel tempo
PSYOPS	psychological operations
QQLR	Quadrennial Quality of Life Review
ROE	rules of engagement

SEAL	Sea-Air-Land team
TAD	temporary additional duty
USAF	United States Air Force
USMC	United States Marine Corps
USS	United States Ship
YOS	years of service

CHAPTER ONE

Introduction

> The armed services are currently under a great deal of stress. . . .
> Without the ability to attract and retain the best men and women, the armed services will not be able to do their job.
> —*Secretary of Defense Donald H. Rumsfeld, "Foreword," The All-Volunteer Force: Thirty Years of Service (Rumsfeld, 2004b)*

In 1970, in the middle of the undeclared wars in Southeast Asia and with America engaged in combat, Congress agreed to President Nixon's proposal to transition to an all-volunteer force. Since then, all branches of the military have relied on volunteers to meet their manpower needs. Currently, with extended deployments in Iraq and Afghanistan, the Army is having difficulties recruiting new non-prior-service personnel. Questions have been raised concerning the viability of the all-volunteer force and how the Department of Defense (DoD) can manage personnel during these times of stress and uncertainty. For example, sociologist Charles Moskos of Northwestern University recently called for a commission to examine the viability of the all-volunteer force that is "independent of the Pentagon" (Bowman, 2005). In addition, the editorial board of the *Dallas Morning News* took recent comments by Chief of the Army Reserve that a decline in recruiting "could provoke a new debate over a draft" (Whittle, 2004) to mean that "a military draft could be around the corner" (*Dallas Morning News* editorial board, 2004).

This report addresses these concerns, with particular attention to the history of conscription and volunteerism. It examines the history of the draft in this country to try to understand when and under what conditions conscription has been used effectively to raise the manpower needed by the Army during wartime. The report also examines what other means besides conscription the Army can use to meet manpower demands. Specifically, what actions can be taken to increase the supply of volunteers or reduce the demand for new personnel?

High retention and a large career force are consequences of 30 years of an all-volunteer force. This puts a premium on retention to both maintain the skills of the Army and keep the number of new recruits at a low level. Today, historically high proportions of military members are married and have children. This report also looks at the many programs that have been developed to help military members and their families cope during difficult times, as well as the particular challenges of understanding which programs work.

CHAPTER TWO

To Draft or Not to Draft, That Is the Question

If there is to be a public debate over conscription, it should consider under what conditions conscription has been used effectively to raise the manpower needed by the Army during wartime. From the earliest period of human history, countries have used both conscription and volunteerism often simultaneously to man their militaries in periods of peace and war. The use of each has been intertwined with ideas of citizenship, sacrifice, efficiency, and effectiveness. At times, conscription has been the norm; at other times, it has been volunteers; and still at other times, both have operated side by side to fill out the ranks.

Conscription Versus Volunteerism—Great Britain, France, and Prussia

The histories of Britain and France are most often used to spotlight the differences between countries that have favored volunteerism and those that have favored conscription and to help illustrate the conditions when conscription has been accepted. In addition, the terms of conscription have changed. Early drafts were not universal and allowed those selected to buy their way out of service. More recently, drafts have been based on the principle of universal obligation and "selective service."

The British Tradition

Great Britain, buttressed by the isolation afforded it by being separated from most adversaries by the sea, was able to provide for the defense of the nation as it limited the power of the state in favor of a military force made up of volunteers. In Great Britain, the 17th-century civil war between king and parliament was fought in part over the power of the king to command the militia. Compelled service was so unpopular that a petition to parliament in 1648 asked members of the Commons to "[d]isclaim yourselves and all future representatives . . . [from using the] power of pressing or forcing any sort of men to service in wars, there being nothing more opposed to freedom" (as quoted in Flynn, 2002, p. 12). As part of the Restoration, the Bill of Rights of 1688 expressly prohibited the king from "raising and keeping a standing army within this kingdom in time of peace unless with consent of Parliament" (Lords Spiritual and Temporal and Commons Assembled at Westminster, 1689). Thereafter, and through the zenith of an expanding British Empire, Britain's power rested largely on volunteers.[1] Even during the Napoleonic period, "the 'ballot' [what we would call the draft today] was reserved as an emergency measure of compulsion in the event that insufficient numbers should volunteer" (Cutler, 1922, p. 12). During the remainder of the 19th century, the British Army engaged extensively in the maintenance of the empire, which resulted in frequent wars and extensive overseas deployments.[2]

---

[1] There still remained a responsibility to serve in the local militia and by "the Ballot Act of 1757, the crown could force men into the militia, then call up this force." For most of its modern history, Britain's small standing and professional Army was made up of volunteers and "functioned mainly in the pacification and policing of the empire" (Flynn, 2002). "Press gangs" were also used to forcibly seize and carry individuals into service. After 1800, England restricted impressments mostly to naval service and generally abandoned such forcible measures after 1835. The system fostered gross abuses and was used to fill the army and navy with a group of men more ready for mutiny, desertion, or other disloyalty than service, and it adversely affected voluntary recruitment. It fell into disuse after 1850. See "Impressment," in *The Columbia Electronic Encyclopedia* (2005).

[2] Kerr notes,

> The British Empire was involved in such conflicts as the Crimean War in Palestine (1853–1856), the Indian Mutiny in India (1857–1858), the Sudan Campaigns in the Republic of Sudan (1885–1897), the Zulu War in South Africa (1879), and the Boer War also in South Africa (1880–1881, 1899–1902). Unfortunately, the once regal and persuasive British Army was spread thin with their involvement in so many ventures all at the same time. . . . During the 19th Century, the British Army was made up of nearly 142,000 men (120,000

The system was based on volunteers even during periods of war,[3] the size of the volunteer Army being a constraint to the foreign policies of Britain rather than a factor to be adjusted through the use of a draft. During the initial years of World War I, Britain "doggedly adhered to the volunteer system of securing a fighting force" (Cutler, 1922, p. 21).[4] In 1916, the enormous manpower demands led to a national conscription. But by March 1920, with occupational duty behind it, Britain ended its draft. It was not until the eve of World War II, and after Hitler had abrogated the Munich Accords that Prime Minister Neville Chamberlain had called "peace in our time," that Great Britain restored conscription (on April 27, 1939).

One of the outcomes of World War II was the demise of the British Empire. Unfortunately, it did not die without a struggle. Between 1946 and 1960, Britain fought six colonial wars. By 1946, it was clear that the manpower needs of the armed forces—for occupation duty in Germany, to support the Greek government, and to preserve law and order in Palestine, as well as for duty in the empire—was such that conscription could not end. The wartime draft law was extended, with

---

> infantry, 10,000 cavalry, 12,000 artillery which included over 600 heavy guns). This impressive number led the military to be one of the more dominating forces in the world at this point in history. Because of their involvement in so many foreign ventures they were forced to limit available men. They had over 32,000 men stationed in Palestine policing the aftermath of the Crimean War while they had another 50,000 men stationed in India looking after ventures there. The remaining 60,000 men were divided among ventures in Africa and homeland security. . . . The British Empire of the 19th Century ultimately tried to acquire too much territory outside the British Islands in too little time. Their imperialistic greed overcame them and forced their military into distress for many decades to come. (Kerr, undated)

[3] A review of the British Army during the Crimean War noted that "[t]he system of recruiting of voluntary enlistments makes it very difficult, in time of war, to keep the efficiency of the army" (*Putnam's Monthly* editorial staff, 1855).

[4] Cutler, comparing the conscription systems of both Germany and France from the Napoleonic period into World War I, notes that "Great Britain adhered to the old methods, a professional army supplemented by volunteers [as a reserve force], well into the World War, although she began vigorously to debate the question of conscription on August 25, 1914; it was not until January 27, 1916, that she overcame her antipathy to the 'ballot' and adopted a selective service law" (Cutler, 1922).

service, first for one year. Then, in 1950, it was extended to two years, with three and one-half years of reserve duty. The Suez Crisis in 1956 led to a reassessment of both the structure of the armed forces and the need for conscription. The Defence White Paper of 1957 argued for an end to conscription and emphasized "the British contribution to the nuclear deterrent and the greater efficiency of the remaining troops in Germany due to better equipment and reorganization" (Whitely, 1987). In April 1957, the British government announced its decision to end conscription. The end was delayed by the erection of the Berlin Wall, and the last conscript was inducted in 1960. By 1963, there were no conscripts serving in the British Army.

The French Tradition

The fundamental difference between Britain and France reflects the difference in philosophy of English philosopher John Locke and French philosopher Jean-Jacques Rousseau, the former emphasizing the rights and liberties of individual citizens and the latter, a citizen's responsibilities to the state.[5] In 1772, Rousseau wrote, "It was the duty of every citizen to serve as a soldier" (as quoted in Flynn, 2002, p. 3). Following that line, in 1789, with foreign powers poised to restore the monarchy, a committee of the National Assembly reported, "Every citizen must be a soldier and every soldier a citizen, or we shall never have a constitution" (as quoted in Flynn, 2002, p. 5). The Declaration of the Rights of Men and Citizens of 1789 stipulated in Article 12 that "[t]he security of the rights of man and of the citizen requires public military forces" and in Article 13 that "common contribution is essential for the maintenance of the public forces" (Representatives of the French People, 1789).

Through conscription, Napoleon built the largest army in Europe to that point in time, ushering in the era of the "mass army." To service the army, the *levée en masse* of 1793 called for 300,000 men between the ages of 18 and 25. The first conscription, however, was the *loi Jourdan* in 1798. It required 20-year-olds to serve for five years. In 1799, the induction age was raised to 22. After the fall of Napoleon,

---

[5] In practical terms, one can speculate that the strategic realities of having a channel of water to offer protection from a land invasion might have also been an important difference.

one of the first things the new king did was to end the draft (in 1814). Conscription returned in 1818 in the form of a national lottery. Service fell, however, only on the unfortunates who had both a low lottery number and no means to pay the 2,000-franc "blood tax" for an "exemption." Moreover, the unlucky conscript had to serve for seven years. This certainly contributed to the humiliation of the French Army by the Prussian Army of conscripts on the battlefield in 1870.[6] In 1872, France moved toward the Prussian system of universal military training, which remained an essential part of the military system until 1996.

After 1989, with the end of the Cold War, and for the first time since 1871, France did not face any threat to its national territory. Unfortunately, conscription and the Cold War army did not lend themselves to the new missions of external actions and multinational operations. In fact, in 1991, following a long tradition of not assigning conscripts outside of metropolitan France without their consent, the French president "vetoed the use of conscripts during Desert Storm operations, [and] France was only able to deploy 12,000 soldiers, from an army of more than 500,000 men" (Irondelle, 2003, p. 162). The two principles of "obligation and universality" on which the draft had been built and which had sustained conscription for 150 years now came into conflict with another principle—equality. With a "structural surplus of people eligible for national service beyond the needs of the military" (Irondelle, 2003, p. 163), fewer and fewer people actually served. In 1991, national service could be accomplished by serving for as little as ten months in the military or by enrolling in one of five forms of civil service—or one could even claim to be a conscientious objector. At a time when France was trying to increase the professionalism of its army by increasing the number of entirely professional units and recruiting more career personnel to restore the legitimacy of the draft, it was also trying to "remilitarize" the draft. The Defense White Paper

---

[6] The French Army was made up of approximately 400,000 regular soldiers. The Prussian Army consisted of conscripts during its initial period of service and reservists who mobilized for the campaign. Through the Krumper system, the smaller Prussian state could field an Army numbering 1.2 million ("Franco-Prussian War," 2005).

of 1994 "reaffirmed the primacy of military service, since 'military service is the raison d'être of conscription'" (Irondelle, 2003, p. 173). Even though action in the former Yugoslavia reinforced the need for a professional army, the civilian leaders of the Ministry of Defense, the service chiefs of staff, and the head of the Joint Chiefs all still favored the continuation of a mixed system of professionals and draftees. It was reported that the head of the Navy felt that conscripts "provid[ed] the Navy with the 'fresh air' it needed," and the Army leadership thought "conscription to be the best guarantee of harmonious relations between Army and society" (Irondelle, 2003, p. 176). With the assumption that a fully professional Army would be a smaller Army, and with fears that the end of conscription would lead to higher unemployment, the closing of bases, and "potential loss of local revenue, . . . conscription appeared [to be safe] for the next twenty years" (Irondelle, 2003, p. 178). In less than a year, however, France started the transition to an all-professional and all-volunteer armed force.

The final move to an all-volunteer force was the result of the election of President Jacques Chirac in 1995 and of a small reform movement that wanted a fully professional military—with "little support within the military institution . . . the idea of professionalization gained ground with young officers who had experienced peacekeeping operations." Chirac was able to push full professionalization of the armed forces through the Defense Council, which "imposed the move to a professional army on the chiefs of staff and the Ministry of Defense" (Irondelle, 2003, p. 183). When the ensuing national debate over the future of national service centered on an even shorter period of service—one proposal being for two months to educate the conscripts and another "a three-month-long national service that ensured military training for all conscripts" (Irondelle, 2003, p. 184)—even the service chiefs of staff knew that it was time to move to an all-volunteer force.

In many ways, the last stage of conscription in France was a replay of the American experience 25 years earlier. Although France was not engaged in an unpopular war, the changing technology of modern warfare, a growing population of draft-eligible young men, the reality that conscription was no longer universal, and the determined leadership of the French president overwhelmed one of the strongest French traditions. As in the United States, France could not quite go all the way in eliminating the draft but maintained a mechanism for a "standby draft,"[7] if the need for a mass army ever arose. But for all practical purposes, it was perceived that the French draft had ended.

Equity and the Prussian Model of Universal Selective Service

The original French model of conscription, with its emphasis on the obligation of all citizens to defend the revolution coexisting with provisions to allow a citizen to buy his way out of service, proved to be a clear contradiction; this model was finally corrected after the humiliating French defeat at the hands of Prussia in 1870. The modern model of universal military service developed by Prussia during and after the Napoleonic period proved so effective in allowing a country to mobilize its manpower and field a much larger army than might have been maintained as a standing force that by the end of the 19th century it was in wide use throughout the non-English-speaking world. Universal military service and a small regular force backed up by reservists who had been trained for several years on active duty provided Germany with a well-trained and large national army available on mobilization to meet the demands of a modern mass army. This model had inadvertently been developed during the Napoleonic War, when Prussia was compelled by the Treaty of Paris of September 1808—after its defeat at the Battle of Jena on October 14, 1906—to limit its army to 42,000 men. The intent of such a limit was to reduce Prussia to a "second-rate" power. Prussia got around this limit by instituting the so-called Krumper system, in which each company sent five men on extended leave every month and took in five recruits so that a trained reserve could be built up over time. By 1813, the Prussian army num-

---

[7] Irondelle notes,

> At the end of the reform process, conscription still was not totally abandoned. The French system preserves a few aspects of conscription. The first is the registration of young men eligible for national military service. This request made by the Defense Ministry was quickly accepted by the political authorities. The military wanted to preserve its ability to mobilize large numbers of conscripts in case a major threat should arise. In fact, the principle of conscription was saved; only military service as such was suspended. (Irondelle, 2003)

bered 270,000—an improvement on the treaty limits of 42,000 (G. S. Ford, 1915, p. 534). With the appointment of Hermann von Boyen as Prussia's "first real minister of war" on June 3, 1814, Prussia was ready to complete the reforms started after the defeat at Jena. On September 3, 1814, Boyen's military law decreed in its opening words that "[e]very citizen is bound to defend his Fatherland," and it established universal military service in Prussia.

Even in light of the new mass armies of the industrial age, the terms of Boyen's law were all encompassing, committing all male citizens from ages 17 to 50 to serve the state. It was the first example of a modern selective service system that countries have subsequently used in time of war—but Prussia used the system even when no war was imminent. Substitutions were banned, but deferments were allowed to maintain essential economic services. Under the law, when a young man turned 20, he was called for five years to the standing army—three years of active duty, followed by two years "on leave" in the reserves. This was followed by seven years in the Landwehr, "with the obligation to serve abroad as well as at home, to participate in occasional reviews and drills on set days, and once annually to participate with the regular army in large maneuvers" (G. S. Ford, 1915, p. 537). A second period of seven years consisted of "occasional drills, the obligation to do garrison duty in war, and possible service abroad in need" (G. S. Ford, 1915, p. 537). Even after 19 years in service, and at age 39, there was a further commitment to the Landsturm until age 50. At the time, critics of the Prussian system argued that it was overly concerned with mass and that Prussia underestimated the amount of military service needed to produce a trained soldier. The French model of long-serving professionals was in vogue, but it had disastrous results for the French.

From the feudal period onward, European countries augmented their professional standing armies with conscripts, whether in the form of the requirement to serve in the militia or more direct forms of conscription—the Prussian system being the most inclusive and systematic application of the concept of universal military training. By the time of Bismarck, all 20-year-olds were called for training, with upwards of 63,000 men entering the army for their period of mandatory training each year. It was this army of short-term conscripts backed by years of compulsory service in the reserves that defeated Denmark in 1864, Austria in 1866, and France in 1870. Thereafter, Austria (in 1868), France (in 1872), Italy (in 1873), Russia (in 1874), and Japan (in 1883) adopted, to one degree or another, the Prussian system of universal military training and selective service (Cutler, 1923, p. 173).

The American Tradition

The noted historian of the modern American draft, George Q. Flynn, suggests that the American tradition is rooted in its colonial past, when military service was seen "less as a part of citizenship and more as a burden imposed by government. Operating under a heritage that stressed minimal government interference with individual choice, these cultures were able to sell military service only as a matter of national defense in an emergency" (Flynn, 2002, p. 3).

Colonial Times

The events in England of the 17th century were all familiar to the American colonists. Ideas concerning service, the role of the militia, and the hostility toward the concept of a standing army "were carried to the English colonies in America where they had a profound impact on the thinking of American leaders" (Schwoerer, 1974, p. 5) and on the Declaration of Independence and the Constitution. The Army's official history of military mobilization notes that in the colonies "every able-bodied man, within prescribed age limits, . . . [was] required by *compulsion* to possess arms, to be carried on muster rolls, to train periodically, and to be mustered into service for military operations whenever necessary" (Kreidberg and Henry, 1955, p. 3, emphasis added). Nevertheless, from 1777 on, the "annual pattern of recruiting" included a congressional allocation of quotas to the states; and through the states to the towns; and, when volunteerism failed, a draft. Charles Royster explains the process:

> The local militia commanders held a muster and called for volunteers. A few men enlisted. Then weeks of dickering started. The state or the town or private individuals or all three sweetened the

> bounty. Meanwhile, citizens who did not want to turn out with the militia were looking for militia substitutes to hire. . . . By the spring or summer, all of the men who were going to enlist that year on any terms had done so, whereupon the state found that it had not filled its quota. . . . Drafting began in 1777 and sent men for terms ending in December, which ensured that the whole process would begin again next January.
>
> Those who enlisted wanted to be paid. After army pay became low, rare and depreciated, these men sought their main compensation in the bounty given at the time of recruitment. . . . When drafting began, it often did not mean selecting an unwilling man to go, but selecting from among the unwilling one man who had to pay one of the willing to go as a substitute. Even then the draftee got a bounty. . . . Apart from the handling of army supplies, recruiting introduced more corruption into American society than any other activity associated with a standing army. . . . Bounties inspired some soldiers to enlist several times with several units within a few days. (Royster, 1979, pp. 65–71)

George Washington saw the draft as a "disagreeable alternative." On January 28, 1778, reacting to the "numerous defects in our present military establishment" and the need for "many reformations and many new arrangements," he wrote to the Committee of Congress with The Army:

> Voluntary inlistments [*sic*] seem to be totally out of the question; all the allurements of the most exorbitant bounties and every other inducement, that could be thought of, have been tried in vain, . . . some other mode must be concerted, and no other presents itself, than that of filling the Regiments by drafts from the Militia. This is a disagreeable alternative, but it is an unavoidable one.
>
> As drafting for the war, or for a term of years, would probably be disgusting and dangerous, perhaps impracticable, I would propose an annual draft of men, without officers, to serve 'till the first day of January, in each year . . . . This method, though not so good as that of obtaining Men for the war, is perhaps the best our circumstances will allow; and as we shall always have an established corps of experienced officers, may answer tolerably well. (Washington, 1931–1944, Vol. 10, p. 366)

On February 26, 1778, Congress acted on the report of the "Committee of Congress at Camp," which had been appointed to work with General Washington in developing recommendations "as shall appear eligible" (Washington, 1931–1944, Vol. 10, p. 362). Congress passed a resolution "that the several states hereafter named be required forthwith to fill up by drafts from their militia, [or in any other way that shall be effectual,] . . . [t]hat all persons drafted, shall serve in the continental battalions [*sic*] of their respective states for the space of nine months (W. C. Ford, 1904–1937, p. 200). The details of the drafting, however, varied among the states, with the common goal of "obtaining recruits with a minimum of governmental coercion" (Royster, 1979, p. 66). By 1781, however, the majority of those who took the field at Yorktown were militiamen.

After the war, in 1783, Washington wrote Alexander Hamilton to endorse the concept of a citizen's obligation to the state: "Every Citizen who enjoys the protection of a free Government, owes not only a proportion of his property, but even of his personal services to the defense of it" (Washington, 1974). The preamble to the American Constitution starts with the words, "We the People," and includes the phrase "provide for the common defense." The Knox plan of 1790 envisioned universal military service. Suspicion of a professional army was one reason for the militia clause of the Constitution (Article I, Section 8).[8] The Second Amendment to the Constitution—the right to bear arms—also reflects the basic suspicion 18th-century Americans had of a professional army and the confidence they had in the militia as the protector of their freedoms. The militia, as stated by the Act of

---

[8] Article 1, Section 8, provides for both the militia and a national Army and Navy. Clauses 15 and 16 provide for the militia: "The Congress shall have Power . . . to provide for calling forth the Militia to execute the Laws of the Union, suppress Insurrections and repel Invasions; to provide for organizing, arming, and disciplining, the Militia, and for governing such Part of them as may be employed in the Service of the United States, reserving to the States respectively, the Appointment of the Officers, and the Authority of training the Militia according to the discipline prescribed by Congress." In addition, Congress is authorized to "raise and support Armies . . . [and in Clauses 12–14 to] provide and maintain a Navy." These latter clauses are also the basis for Congress's power to order a national conscription. (See the discussion of Arver of January 7, 1918, in Chambers, 1987, p. 219, and the Selective Draft law cases in O'Sullivan and Meckler, 1974, pp. 140–149.)

1792, was to be made up of "each and every free able-bodied white male ... [between] the age of eighteen years and under the age of forty-five years" (O'Sullivan and Meckler, 1974, pp. 36–37). Nevertheless, when Alexis de Tocqueville traveled through the United States in the 1830s, he noted, "In America conscription is unknown and men are induced to enlist by bounties. The notions and habits of the people of the United States are so opposed to compulsory recruiting that I do not think it can ever be sanctioned by the laws" (de Tocqueville, 1835, Book 1, Chapter 13).

From the Revolution to the Civil War

Even though, as mentioned above, the Militia Act of 1792 mandated conscription, by the time of the Mexican War (1846–1948), service in the militia had ceased to be compulsory (Cutler, 1922, p. 41).[9] The small professional army of the federal government was never designed to do more than maintain the military infrastructure of the nation and to provide a core that the militia and those who voluntarily answered "the call to the colors" could build on.[10] The tremendous manpower demands of the Civil War changed that and resulted in America's first draft.

---

[9] Cutler notes, "Between 1815 and 1846, the years of Jacksonian democracy, militia service was everywhere allowed to become voluntary; the law of the United States was tacitly annulled by the states. Volunteer companies . . . sprang up in large numbers as substitutes for the older force. . . . Congress decreed that they should be regarded as militia and organized under the militia clause of the Constitution" (Cutler, 1923).

[10] Cutler (1922) notes that, after the Revolution, on July 2, 1784, the Army was disbanded with the exception of one company of soldiers that was retained to protect the military stores of the Nation at West Point and Fort Pitt. By 1798, the Army totaled 2,100. At the start of the War of 1812, about 80,000 volunteers and militia augmented the regular Army of 6,744. At the start of the Mexican War, the regular Army was 8,349, and at the start of the Civil War, the regular Army numbered 16,367.

The term "call to the colors" is technically a bugle call to render honors to the nation. It is used when no band is available to render honors, or in ceremonies requiring honors to the nation more than once. "To the color" commands all the same courtesies as the national anthem. The sound of the bugle call can be heard at http://www.globalsecurity.org/military/systems/ground/images/11colors.mp3 (as of June 2006).

Ironically, the first American Congress to pass a "national" conscription law was the Congress of the Confederate States of America.[11] On April 16, 1862, the Confederate Congress provided that every able-bodied white male between the ages of 18 and 35 serve in the army for three years (Cutler, 1922, p. 83). It also extended the enlistments of those who had already volunteered for the duration of the war. The draft was unpopular, and only 21 percent of Confederate soldiers were conscripts. By one account, "[T]he Rebel soldiers hated the Conscript Law. It was unfair, and they knew it. It took the glory out of the war, and the war was never the same for them." A Confederate general summed up the situation: "It would require the whole army to enforce conscription law, if the same thing exists through the Confederacy which I know to be the case in Georgia and Alabama, and Tennessee" (as quoted in Cutler, 1922, p. 68). But was this not to be expected? Albert Moore in his definitive account of *Conscription and Conflict in the Confederacy* saw the difficulties the South faced as being

> inherent in a system of compulsory service among a proud and free people. Conscription was not only contrary to the spirit of the people but to the genius of the Confederate political system. It seemed unnatural that the new government, just set up as the agent of the sovereign States, should exercise such compelling and far-reaching authority over the people, independent of the States. . . . Conflict with State authorities in the enforcement of it—conscription—seriously impaired its efficiency. (Moore, 1924, p. 354)

Nevertheless, conscription in the South fared far better than in the North, and throughout the war it provided the manpower the South needed to carry on the fight.

At the start of the Civil War, President Abraham Lincoln called for 75,000 militiamen and volunteers, the former to serve for a matter of months and the latter for one or two years (Chambers, 1987,

---

[11] In 1814, those arguing that it was the right of the states to raise the militia had blocked President Madison's proposal for a national draft. Nearly a half-century later, in 1862, it was the "Confederate Congress [that] threw the theory of states' rights to the winds and enacted the first 'Conscription Law'" (Cutler, 1923).

p. 42). Eventually, though, it was clear that something more than the militia was required. On March 3, 1863, Lincoln signed the Union's first draft law, the Enrollment Act (O'Sullivan and Meckler, 1974, pp. 63–66). Notably, the Enrollment Act made no mention of the militia and asserted, for the first time, the federal government's authority to directly draft people into the national army.[12] The draft, however, was a despised institution because there was little sense of equal sacrifice. Following the French tradition, wealthy men were able to buy their way out of service—commutation—or hire a substitute to serve in their stead.[13] The draft riots in Boston, New York, and other Northern cities attested to the draft's unpopularity.[14] In the most perverse way, the draft was effective in the North, not because it brought in large numbers of people, but because it persuaded "elected officials to raise much higher bounties to entice men to enlist and thus avert the need for governmental coercion" (Chambers, 1987, p. 64). Local bounties soared to as high as $1,500 (Kreidberg and Henry, 1955, p. 110). By the end of the Civil War, states and localities had paid almost a quarter of a billion dollars in bounties to encourage young men to volunteer, with the federal government spending an amount only slightly greater. By one estimate, "Bounties cost about as much as the pay for the Army during the entire war . . . and five times the ordnance costs" (Kreidberg and Henry, 1955, p. 110). In fact, the bounty program became so popular that many men volunteered again and again. One "bounty jumper"—as such men were known—was reported to have enlisted 32 times (Cutler, 1922, p. 64). By the end of the Civil War, 2.1 million men saw service in the blue uniform of the Union. Of this number, the

---

[12] The act provided those drafted would "remain in service for three years or the war, whichever ended first" (Kreidberg and Henry, 1955).

[13] O'Sullivan and Meckler note, "At the heart of the antagonism to the draft law lay the realization that the commutation fee of $300, not to mention the possibility of hiring a substitute, was far beyond the means of most workmen" (O'Sullivan and Meckler, 1974).

[14] On July 2, 1863, the second day of the battle of Gettysburg, President Lincoln called for 300,000 men to be drafted; 20 percent of those enrolled. In New York City, names were drawn on July 11. The next day, the newspapers printed both the names of those drafted and the lists of those killed at Gettysburg. The next morning, riots broke out that lasted three days.

draft produced 46,000 conscripts and 116,000 substitutes, and 87,000 paid the commutation fee to buy their way out of service. The rest were volunteers.

From the Civil War to World War I

One lasting legacy of the Civil War was the *Report on the Draft in Illinois* (see Oakes' report in O'Sullivan and Meckler, 1974, pp. 93–101), prepared in 1865 by the Acting Assistant Provost Marshal General of the State of Illinois, Brevet Brigadier General James Oakes. This report became the blueprint of the next draft, which was not until World War I. Between the Civil War and World War I, including the Indian Wars and the Spanish-American War, America relied on volunteerism and the *new* volunteer militia of the states—the National Guard—to provide the manpower needed to defend the country. In 1915, with war raging in Europe and a growing preparedness movement, the idea of universal military training and service, already popular in Europe, was being widely discussed. President Woodrow Wilson still did not want a large standing army. In his State of the Union address that year, he told Congress, "Our military peace establishment . . . [should be] no larger than is actually and continuously needed for the use of days in which no enemies move against us" (Wilson, 1915). He did, however, call on Congress to approve an increase in the standing army of some 31 percent, to 141,848, and a new force of "disciplined citizens, raised in increments of one hundred thirty-three thousand a year through a period of three years . . . for periods of short training for three years [not to] exceed two-months [*sic*] in the year" to supplement the army. Even though he wanted this force to make "the country ready to assert some part of its real power promptly and upon a large scale, should occasion arise," he did not want a draft. While he saw "preparation for defense as . . . absolutely imperative," he wanted to "depend upon the patriotic feeling of the younger men of the country whether they responded to such a call to service or not" (Wilson, 1915).[15]

---

[15] The National Defense Act of June 3, 1916, actually raised the standing army to 175,000 and provided for a reserve of 450,000 (O'Sullivan and Meckler, 1974).

Wilson's views about a draft changed as it became increasingly clear that America would enter the war. On April 2, 1917, he asked Congress for a Declaration of War. Four days later, the day Congress actually declared war on Germany, the President asked for a draft, and on May 18, 1917, he signed the Selective Service Act of 1917 into law. Unlike the Civil War draft, the new draft was widely accepted. Frederick Morse Cutler described the "marvelously complete response; . . . the popular support and approval accorded the selective service" and how on the day young men reported for registration, "a feeling of solemnity possessed all hearts; a holiday was declared; at the stated hour, church bells rang as though summoning men to worship" (Cutler, 1923, p. 174).[16]

The new Selective Service law provided that both draftees and enlistees serve for the duration of the war and that compulsory military service should cease four months after a proclamation of peace by the President. Although the law did not allow for bounties or personal substitution, it did provide for deferments based on essential work. The term Selective Service was used to capture the idea that, while all men of a specific age group—eventually 18 to 45 years of age—might be required to register, only some would be selected for military service in line with the total needs of the nation. For example, the approximately 2.8 million men who were drafted during World War I were only a fraction of the 23.9 million men registered and classified. The 72 percent of the armed forces who were draftees made a better case for equality of sacrifice than did those drafted during the Civil War (Chambers, 1991). Even then, when the needs of the mass army ended, so did the need and legitimacy of the draft.[17] While there was some interest in

---

[16] Cutler's account of popular support for World War I is in sharp contrast to Mueller's assessment that, retrospectively, World War I "was the most unpopular war of the [20th] century." His account, of course, is based on responses to a question asked in 1937, when 72 percent of those responding thought it had been a mistake to go to war in 1916. This really illustrates the fact that support for wars varies over time. In 1971, Mueller wrote that World War II presumably was the most popular in American history (Mueller, 1971). He notes, however, that depending on when the question was asked and the specific wording of the question, different conclusions could be reached.

[17] Chambers notes, "What was most significant about the draft in the immediate post–World

retaining some form of involuntary military training after World War I (see General Leonard Wood's call for universal military training in O'Sullivan and Meckler, 1974, pp. 117–120), peacetime conscription, limited budgets, and a relatively small standing army could not lay claim to the compelling argument of "equal sacrifice" that had been so successfully used at the beginning of World War I. It was not until 1940, months after the start of the Second World War in Europe, that the conditions were again right for Congress to vote for a draft.

World War II

From the end of World War I until 1926, no work was done on the future mobilization of manpower for the armed forces. In 1920, through the efforts of a number of people who had participated in the wartime Selective Service System, the National Defense Act gave authority for "mobilization of the manhood of the Nation . . . in an emergency" (as quoted in Hershey, 1942) to the War Department General Staff. It took six years before the Secretaries of War and the Navy created the Joint Army-Navy Selective Service Committee (JANSSC). In 1936, the "entire operation consisted of two officers and two clerks" (Flynn, 1985, p. 63), when Army Major Lewis B. Hershey was assigned as the executive. Hershey got the assignment because of his reputation of being a good staff officer and his "talents at management and personnel [and] . . . Hershey had originally come from the National Guard, an outfit which had to play a big role in the conscription plan" (Flynn, 1985, p. 63). Under Hershey's leadership, the JANSSC got an annual allocation of $10,000. He brought in National Guard officers and started to promote training through a number of conferences held throughout the United States. After Congress authorized the draft, the JANSSC operated as the national headquarters of the newly authorized Selective Service System.

War I period is how quickly America abandoned it. . . . By the spring of 1920 Congress had rejected any kind of compulsory military training in peacetime and reduced the wartime army of nearly 4,000,000 citizen-soldiers to [a volunteer] force that numbered only 200,000 regulars" (Chambers, 1987).

On September 16, 1940, President Franklin D. Roosevelt signed the Selective Training and Service Act of 1940, the first peacetime conscription law in the history of the United States. One month later, on October 16, 1940, all men between the ages of 21 and 36 were required to register. A national lottery was held on October 1, 1940, to establish the order of call; with Roosevelt looking on, Secretary of War Henry Stimson drew the first number (Flynn, 1993, p. 22). On November 8, 1940, Roosevelt ordered that no more than 800,000 men be selected and inducted by July 1, 1941 (Hershey, 1942, p. 27). The prescribed period of active service was for one year, to be followed by ten years in the reserves. On June 28, 1941, the president ordered that during FY 1942 an additional 900,000 men would be "selected and inducted." On August 19, 1941, by only one vote in the House of Representatives—and some say by the employment of a quick gavel by Speaker of the House Sam Rayburn—Congress extended the period of service to 18 months by passing the Service Extension Act of 1941 (the debate during the summer of 1941 is described in Flynn, 1993, pp. 51–52). It also reduced the age a person might be inducted to 28 years of age, allowing some 193,000 to leave service before their training period had been completed.

The Principle of Equal Sacrifice. When President Roosevelt signed the draft bill in 1940, he talked about the "duties, obligations and responsibilities of equal service" (as quoted in Flynn, 1993, p. 2). In the preamble of the act, Congress declared:

> In a free Society the obligation and privileges of military training and service should be shared generally in accordance with a fair and just system of selective compulsory military training and service. (As quoted in Hershey, 1942, p. 33)

Using the Selective Service model first introduced in America during the World War I draft, deferments were provided for government officials and for those "employed in industry, agriculture or other occupations or employments" that were "necessary to the maintenance of the public health, interest and safety" (Hershey, 1942, p. 35). The law prohibited deferments for "individuals by occupational groups or of groups of individuals in any plant or institutions" (Hershey, 1942,

p. 37). Given, as one historian noted, that "the draft had been sold as a democratic mechanism" (Flynn, 1993, p. 41), students were only allowed to complete the academic year they were currently in. The "importance of universality of service as befitting a democracy" and "social and economic realities" of the nation was also tested when it came to married men and fathers. While not specifically identifying these two classes, the law allowed the president to defer "those men in a status with respect to persons dependent upon them for support which renders their deferment advisable" (Hershey, 1942, p. 35).

The Importance of Families. As they would later be with the all-volunteer force, issues of family were important to manpower planners even in 1940. In the report to the President on the operation of the peacetime draft, the director of Selective Service noted his special responsibility to "preserve the family life of the Nation intact, to the greatest extent possible." He wrote:

> It was a special objective of Selective Service to preserve the family life . . . even though the fundamental problem which Selective Service dealt with was not marriage, but dependency. . . . The dependents must, in fact, depend on the registrant's income for his support, but while the financial problem was the main, the spiritual relationship and dependence was not to be entirely disregarded. . . . Generally speaking, the deferments on grounds of dependency were generous and were warranted by the peacetime situation in which the decisions were made. (Hershey, 1942, p. 137)

In the final analysis, however, with 10 million of the 17 million men (Hershey, 1942, p. 143) who initially registered for the draft receiving dependency deferments, this effort "to preserve the family life of the Nation" would conflict with both the manpower needs of the war effort and the underlying concept of the universality of service. A Gallup poll four months after Pearl Harbor found that 71 percent of the nation favored drafting men without regard to dependence "if it would be necessary to win the war" (as cited in Flynn, 1993, p. 70). Congress was loath to draft fathers. Even after the passage of the Servicemen's Dependents Allowance Act on June 23, 1942, which made the government responsible to provide support to the wives and children of enlisted men, Congress restricted the drafting of fathers until "all other eligible men were taken" (Flynn, 1993, p. 74). Eventually, with the manpower needs of the war effort expanding and the pool of eligible men running dry, deferments decreased—by 1943, deferments for dependency shrank to 8 million, and by 1945, they were less than 100,000. Nevertheless, the impact that the war had on families remained contentious to include policies on sole surviving sons as portrayed in the film *Saving Private Ryan*.[18] For the most part, however, draftees were initially inducted for the duration of the war plus six months, and most of those assigned overseas did not have the opportunity to return home until the end of the war. After the fall of Germany, the draft was extended for one year to cover the continuing war with Japan and occupation duty in Europe. The original "duration plus six months" notwithstanding, service members were mustered out based on age and points earned for service, which were heavily weighted for combat, overseas service, and paternity.[19]

The Cold War Draft: 1947–1973

Even before the end of World War II, but with victory clearly ahead, Congress, under considerable pressure from the public, pressed the new Truman administration to end the draft. It made little difference

---

[18] The provisions for sole serving son or brother are discussed in Powers (2005). Also see Assistant Secretary of Defense (Force Management and Personnel) (2003).

[19] The Adjusted Service Rating was first calculated as of May 12, 1945. It provided one point for each month of service since September 1, 1940; one point for each month of service overseas; five points for each combat decoration; and 12 points for each child under age 18 to a maximum of three (Jeffcott, 1955). According to Wiltse (1955),

> Officers as well as enlisted men who possessed the critical score could no longer be held in the Army on the ground of military necessity, except in special instances. Adjusted service ratings were recomputed as of 2 September 1945. The critical score of enlisted men was then reduced from 85 to 80 points, and enlisted men 35 years of age and over who had had at least 2 years' service were ordered released on their application; the age for automatic release of those with less than 2 years' service remained at 38, having been reduced from 40 earlier. Within the next 3 months, the critical score for enlisted men was brought down by successive cuts from 80 to 55, while new alternatives of 4 years' service or the possession of three dependent children also qualified men for discharge.

that America faced a sizable need for military manpower to meet the new occupation requirements. Flynn notes, "The public's position on the draft seemed clear: bring the troops home and immediately and stop taking boys through the draft" (Flynn, 1993, p. 89). For President Truman, reviving the notion of universal military training, which had not taken hold after World War I, was the best way to have sufficient manpower to meet the needs of the occupation and to forestall the call to end the wartime draft. Unable to move either the public or Congress to accept universal military training, Truman agreed to end the draft on March 31, 1947. In less than a year, however, the world situation had so deteriorated and the Army's experience with this version of an all-volunteer force had been so disastrous—with a requirement of 30,000 recruits a month, only 12,000 volunteers were coming forward—that Truman asked for a resumption of the draft.

The reinstatement of the draft in the spring of 1948 seemed imperative. The communist coup in Czechoslovakia in February 1948 and the March 5 warning by General Lucius Clay, U.S. military governor in occupied Germany, of the possibility of imminent conflict with the Soviet Union led to President Truman's call for the "temporary reenactment of selective service" (as quoted in Friedberg, 2000, p. 174–175). The "danger," however, passed and the call-up lasted for only three months. By February 1949, inductions were suspended, and by the summer of 1949, the Associated Press reported that "unless an unforeseen emergency develops, the peacetime draft of manpower for the armed forces is expected to expire June 25, 1950" (Associated Press, 1949). On June 24, 1950, North Korean forces invaded South Korea. Three days later, Congress voted for an extension of military conscription.

The Korean War, and the war in Vietnam a decade later, did not mobilize and unite the country as the two World Wars had done or at least had initially done.[20] This was a new kind of undeclared and limited war. This was a war not to achieve unconditional victory but to contain communism. Rather than conscript millions for the duration

[20] Mueller notes that World War II was "unquestionably much more highly supported by the public than the Korean and Vietnam wars" (Mueller, 1973).

of the conflict, the Cold War draftee served two years on active duty, followed by service in the Selected Reserves. The transition from peace to war and back again to peace could be seen not in the mass mobilization of the country but in increases and decreases of the monthly draft calls. When the Korean armistice was signed, American troops remained in Korea and the draft stayed in place. With the perceived threat from communism remaining, the only way one could tell that the war was over was from the monthly quotas assigned to Selective Service, which were getting smaller. In truth, this was not the end of a war but the end of a battle. The Cold War and draft continued, and the nation settled into what Friedberg has called "a state of equilibrium." It was this equilibrium, he argued, that was "less sturdy and less stable than it appeared." He noted:

> Limited conscription—from Korea to Vietnam—aroused little opposition so long as the number of those drafted remained relatively small, the use to which they were put retained broad public approval, those who preferred to avoid service could do so with relative ease, and the inevitable inquiry of the selection process did not receive undue attention. If one of these parameters changed, support for the draft would weaken: if all of them changed at once it would disappear altogether. (Friedberg, 2000, p. 179)

The title of one of numerous government studies captured the enduring problem of the Cold War draft, *In Pursuit of Equity: Who Serves When Not All Serve* (Marshall, 1967). When President Kennedy took the oath of office on January 20, 1961, peacetime conscription had "become the new American tradition" (O'Sullivan and Meckler, 1974, p. 220).[21] If it was a tradition, however, it was one that had been in place for only eight out of the then 185 years of the nation's history, and one that did not affect most Americans. Monthly draft calls were low.[22] For about a decade, from the end of the Korean War to

---

[21] In the words of one author, "John Kennedy primed the pump [that would eventually lead to Vietnam and the end of conscription]. . . . He proclaimed the United States willing to 'pay any price, bear any burden, meet any hardship' to advance the cause of freedom around the world" (Timberg, 1995).

[22] O'Sullivan argued that, because of deferments in the early 1960s, the burden of service

the beginning of the war in Southeast Asia, the Cold War draft was essentially voluntary. While there were high-profile cases, such as the drafting of Elvis Presley, deferments had become more common than inductions. LTG Lewis Hershey, the Director of Selective Service, put it this way: "We deferred practically everybody. If they had a reason, we preferred it. But if they didn't, we made them hunt one" (as cited in Flynn, 1985, p. 218). Later, Hershey would admit that "equity was unattainable" and that "we defer people . . . because we can't use them all" (as cited in Flynn, 1985, p. 225).

The zenith of support for Selective Service came the spring of 1963. On March 5, 1963, the House of Representatives voted 387 to 3 to extend induction authority. The Senate approved the extension by voice vote on March 15. President John F. Kennedy signed the extension of the draft into law on March 28, 1963 (Flynn, 1985, p. 222). However, with low draft calls and an increased pool of draft-eligible men, the anomaly of "a draft agency that did more deferring than drafting" (Flynn, 1985, p. 218) raised questions concerning the equity of conscription and the policies being followed by Selective Service.[23] The noted military sociologist James Burk found that

> the perception of inequities eroded public confidence in the draft. In 1966, for the first time since the question was asked, less than a majority (only 43 percent) believed that the draft was handled

---

fell on the "least vocal and least powerful group in society" (O'Sullivan and Meckler, 1974). However, Janowitz found that "those whose education ranged from having completed nine years of school to those who had completed college, roughly the same proportion [about 70 percent] had had military service (Janowitz, "The Logic of National Service," in Tax, 1967).

[23] Specifically, Walter Oi described the situation this way: "By 1963, . . . the size of the draft-eligible pool had increased relative to manpower demand. To equilibrate supply with demand, draft boards became more lenient in granting deferments and raised mental and physical qualification standards to reduce the size of the I-A classification [highly qualified] pool. The law prescribed an 'oldest first' rule for selecting qualified youths from the I-A pool. Hence, a decline in the number of draftees raised the average age of inductees. The loss of civilian earnings and disruption of career were more costly to older draftees, who also were more likely to voice their objections to involuntary military service. The discretion practiced by autonomous local draft boards . . . led to inequities" (Oi, "Historical Perspectives on the All-Volunteer Force: The Rochester Connection," in Gilroy et al., 1996).

> fairly in their community. Although the public still supported the draft, the problems protesters exposed raised serious questions about its operation during the Vietnam War. (Burk, 2001)

Burk's observations on inequities and public confidence echoed those of Alexis de Tocqueville more than a century before, when he wrote: "The government may do almost whatever it pleases, provided it appeals to the whole community at once; it is the unequal distribution of the weight, not the weight itself, that commonly occasions resistance" (de Tocqueville, 1835, Chapter 23).

The End of Conscription and the Beginning of the All-Volunteer Force

On October 17, 1968, at the height of the Vietnam War, the Republican candidate for president, Richard Nixon, addressed the nation on the subject of conscription. In that speech, he put himself squarely on the side of an all-volunteer force:

> I feel this way: a system of compulsory service that arbitrarily selects some and not others simply cannot be squared with our whole concept of liberty, justice and equality under the law. Its only justification is compelling necessity. . . . Some say we should tinker with the present system, patching up an inequity here and there. I favor this too, but only for the short term. But in the long run, the only way to stop the inequities is to stop using the system. (Nixon, 1968)

One week after taking office, Nixon told his Secretary of Defense, Melvin Laird, to "begin immediately to plan a special Commission to develop a detailed plan of action for ending the draft" (Nixon, 1969). On February 21, 1970, the Commission on an All-Volunteer Armed Force (known as the Gates Commission) forwarded its recommendation to President Nixon to end conscription. The commission unanimously found that the cost of an all-volunteer force was "a necessary price of defending our peace and security . . . [and that conscription] was intolerable when there is an alternative consistent with our basic national values" (Gates, 1970, p. 10). On September 28, 1971, President Nixon signed Public Law 92-129, ushering in the era of the all-volunteer force.[24]

Arguments for an All-Volunteer Force. To some people, including Senator Sam Nunn (D-Ga.), an influential politician at the time, the recommendation of the Gates Commission and the subsequent actions by President Nixon and Congress seemed to be "the clear results of the [unpopular] Vietnam War" (Office of Senator Sam Nunn, 1973); however, others believed that there was a "rational, intellectual basis for the volunteer force" (Lee and Parker, 1977, pp. 524–526).

The arguments the Gates Commission put forward to support its recommendation that the United States give up conscription stand in sharp contrast to the prevailing philosophy in Europe, captured in the preamble of Hermann von Boyen's military law: "Every citizen is bound to defend his Fatherland" (G. S. Ford, 1915, p. 537). The Gates Commission argued:

> The United States has relied throughout its history on a voluntary armed force except during major wars and since 1948. A return to an all-volunteer force will strengthen our freedoms. . . . It is the system for maintaining standing forces that minimizes government interference with the freedom of the individual to determine his own life in accord with his values. (Gates, 1970)

The commission's recommendation to move to an all-volunteer force echoed the arguments that had been heard at the University of Chicago Conference on the Draft in 1966 (Tax, 1967). First, as the commission put it, "conscription is a tax," and it found the tax to be inequitable and regressive. The commission argued that a full account-

---

[24] The law actually authorized "an extension of the draft for two years until July 1, 1973. It increased military pay a total of $1.8 billion over nine months. The largest increase was in basic pay primarily for those with short service ($1.4 billion). Other increases included basic allowance for quarters ($305 million) and dependent assistance allowance ($120 million). . . . Enlistment bonuses were authorized up to $6,000 for men enlisting in the combat elements, . . . initial use of the authority, . . . $3,000." The bill "restores to the President discretionary authority which he had before the 1967 Selective Service amendments, over student deferments and establishing a uniform national call" (White House Press Secretary, 1971).

ing for the true cost of the draft meant that, even given the higher budget costs of an all-volunteer force, a mixed system of volunteers and conscripts was more costly to society than an all-volunteer force. Second, by not accounting for the true cost of the labor employed by DoD, the armed forces were "inefficient" and were wasting society's resources.

The role of the conscription "tax" in arguing for an all-volunteer force was so central to the Gates Commission's conclusion that the commission devoted a whole chapter (Chapter 3) to presenting its argument. The commission invoked Benjamin Franklin's writings on the impressing of American sailors to ask if it was "just . . . that the richer . . . should compel the poorer to fight for them and their properties for such wages as they think fit to allow, and punish them if they refuse?" The importance of this argument was highlighted, as the final report noted:

> This shift in tax burden lies at the heart of resistance on "cost" grounds to an all-volunteer armed force. Indeed, this shift in tax burden explains how conscription gets enacted in the first place. In a political democracy conscription offers the general public an opportunity to impose a disproportionate share of defense costs on a minority of the population. (Gates, 1970, p. 25)

Fair Pay and Conscription. When the Congress debated the end of conscription in 1970, the fate of the Cold War draft was very much uncertain. The issue made strange bedfellows. Some liberals in Congress, such as Senator Edward Kennedy (D-Mass.), argued that "a volunteer force during wartime would be mercenary, composed of the poor, black, and uneducated" (as quoted in Lee and Parker, 1977, p. 96). Some conservatives, such as Senator John Stennis (D.-Miss.), thought that a volunteer force was "a flight from reality." On the other side of the issue, liberals such as Senator Mike Mansfield (D.-Mont.) and conservatives such as Senator Barry Goldwater (R.-Ariz.) found common ground in supporting the abolition of the draft. All could agree, however, that pay should be fair, and, as pay rose, so did the number of young men who volunteered. The end of the draft was certain when it became clear at market wages that there would be enough volunteers to man the force.

Bring Back the Draft: 1981 and 2004

In 1981, the Army, anticipating a decline in voluntary enlistments, "proposed a five-year calling for reinstitution of the draft by the fall of 1984 to meet its goal of expanding the active force by about 100,000 over the next five years" (Hardyman, 1988). Secretary of Defense Casper Weinberger rejected all moves to return to the draft, telling the *New York Times*, "We know what the draft did to the social fabric of this country in the '60s" (as quoted in Chambers, 1987, p. 259).

In 2004, with the all-volunteer force under hostile fire and strained by long deployments and with a vote pending in the House of Representatives to bring back the draft, Secretary of Defense Donald Rumsfeld wrote to the Chairman of the House Armed Services Committee his assessment of the all-volunteer force and his thoughts on conscription, noting:

> A draft simply is not needed. We have 295 million people in the United States of America and there are some 2.6 million active and reserve forces serving. We are capable of attracting and retaining the people we need, through the proper use of pay and other incentives. . . .
>
> In danger zones across the globe, the all-volunteer, professional force is performing superbly—as typified by operations in Afghanistan and Iraq. I have met with many of these men and women as they carry out their missions. They are committed, enthusiastic, and proud to be contributing to the defense of the nation. Most importantly, they want to be doing what they are doing. Every single one of them stepped forward, raised their hand, and said, "I'm ready. I want to serve." They are serving most professionally and proudly. (Rumsfeld, 2004c)

Shortly afterward, the House voted 400 to 2 to reject a return to conscription (Babington and Oldenburg, 2004).

What History Tells Us

Until the Civil War, the United States had relied on a small professional army augmented in times of national emergency by the militia and volunteers. In times of war, from the Civil War up to today, the United States has used conscription four times. The draft was successful in meeting the manpower needs of the country twice, and twice volunteerism effectively replaced it. Conscription was successful during the two World Wars, when the conflict had general popular support,[25] the entire male population of military age was included (registered), and selection was judged to be fair and sacrifice perceived to be equal—equal in terms of the chance to serve, not in terms of the economic consequences of serving, or as the preamble of the 1940 draft law put it, "shared generally in accordance with a fair and just system of selective compulsory military training and service."[26]

Conscription was unsuccessful during the Civil War and the Cold War, when the cause did not enjoy the full support of the people; it was also unsuccessful when selection appeared to be random or biased with inequitable service, which was apparent in both the Civil War and the Vietnam War.

---

[25] Popular support at the time conscription was instituted. Mueller's research on the two World Wars, Korea, and Vietnam clearly shows the difficulties in measuring both popularity and how it changes over time. Cutler describes the "marvelously complete response; . . . the popular support and approval accorded the selective service" (Cutler, 1923), which stands in sharp contrast to Mueller's contentions about the retrospective views about World War I (Mueller, 1971). For the purposes of the draft, it is Cutler's version of popular support that counts.

[26] Economists would argue that conscription by definition implies unequal sacrifice. If a government wanted to pay market wages to attract soldiers, then it would not need to draft them at all. Alternatively, if the government wants to maintain an army with submarket wages, then it needs to use force and require participation of all or substantial portions of the population. By construction, if there is national support for "the cause," then the population of draftees is not sharing equally in the cost of military service when they are not paid their market wages. Alternatively, political scientists and sociologists look at the term "fair" in terms of equality of services without regard to the market. For them, citizens should share equally in national defense, regardless of their economic situation. Selective service that channels people into needed occupations reflects the perceived value of one's service to society. In an ideal situation, the value to society should be reflected in the market wage, and the private and public solutions would be identical.

Particularly interesting about the notion of support for the draft being associated with support for the cause and a sense of universal sacrifice is that it often runs counter to the *self-interest* of those involved. One might expect that, with so much at stake, including the very life of those being drafted, opinion about war and peace and views on conscription would be increasingly negative the more likely it is that one may be asked to serve. However, this may not to be the case, as John Mueller found in his 1973 study: "On the whole, the data do not suggest that self-interest is a very good predictor of hawkish or dovish attitudes." Citing Barton's 1968 study of students at Columbia University, Mueller went on to note that "support for American withdrawal from Vietnam was completely unrelated to the draft status of the respondent" (Mueller, 1973, p. 149).

One might also ask whether the level of support affects the viability of the draft. In 1968, presidential candidate Richard Nixon called for the end of conscription, noting:

> All across our country we face a crisis of confidence. Nowhere is it more acute than among our young people. They recognize the draft as an infringement on their liberty, which it is. To them, it represents a government insensitive to their rights, a government callous to their status as free men. They ask for justice, and they deserve it. (Nixon, 1968)

Within days of assuming the presidency, Nixon took the first steps to move to an all-volunteer force. It is important to note that Nixon's stance took place at a time when a majority of Americans apparently still favored the draft. That same month, 62 percent of those responding to a Gallup poll thought that "after the Vietnam war is over, . . . the U.S. should [not] do away with the draft and [should not] depend upon a professional military force made up of volunteers." The majority of respondents thought that "the draft should be continued" (Gallup Brain, 2006a). Thus, it would appear that, strictly speaking, when it comes to the draft, the majority does not always rule.

History suggests that a sense of fairness is also important in achieving a level of support for the draft, and in this regard there may be no such thing as a viable "little draft." A "little draft" can never be equitable, and a draft to augment volunteers through a lottery, even if it is a random lottery, can hardly provide equality of service needed to gain the support of the country. Sacrificing for one's country is not the same as going to Las Vegas. While the phenomenon of "gambler's remorse" may be a private matter for private wagers, unequal sacrifice resulting from having a low lottery number in the draft can hardly sustain conscription as a viable institution.

American history suggests that conscription works only when (1) the cause enjoys overwhelming support among the general population where even small but vocal and intense opposition is enough to compromise the viability of a conscription, and (2) there is a generally held belief that all are participating with equal sacrifice. Without both of these conditions in place, conscription has not been and will not be a viable way to raise the manpower needed by the military.

Are Conditions Right Today for a Return to Conscription?

In light of the question posed by the *Dallas Morning News* editorial board about whether a military draft could be just around the corner, it seems that the dual requirements of overwhelming support and equal sacrifice that must exist before the American people will accept a draft are not present today.

Support. Before the invasion of Iraq, the nation was almost equally divided over whether to go to war with Iraq. For a six-month period starting with the fall of Baghdad in April 2003, more than 60 percent of the public answered "yes" to the question "Is/was the war worth it?" (Voeten and Brewer, 2004). Since the spring of 2004, a slight majority has answered "no" to the question. In addition, in the fall of 2004, an overwhelming majority (85 percent) told the Gallup polling organization "no" to the question "Do you think the United States should return to a military draft at this time, or not?" (Gallup Brain, 2006b). It would appear, then, that the current conflict does not enjoy the overwhelming support needed to bring back the draft.

Equal Sacrifice. Even if the military is not able to retain sufficient numbers of people to meet all its future requirements, it is unlikely that the numbers of men who would need to be drafted would be so large as to meet the criterion of "equal sacrifice" for the draft to be judged equitable. For example, one might question whether the *Dallas Morning News* editorial board would have written "Our Guard Is Down" (2004), if it knew that for the first three months in FY 2005 the Army Reserve missed its recruiting goal by only 315 recruits (Whittle, 2004). The current shortfalls in the Army and Air National Guard are somewhat larger, but given the size of any draft-eligible pool,[27] conscription would hardly result in anything like equality of sacrifice between those drafted and those not drafted.

Clearly, the military can increase the supply of new personnel by offering them any number of incentives to enlist, and it can try to reduce the need for new personnel. It can also try to increase retention of those already in the force by helping military families cope with the demands of such things as long and repeated deployments. The next chapter deals with these policy areas.

---

[27] For example, with approximately 2 million young men registering for the draft each year, and with approximately half of them unfit for service, the historical rejection rate, upwards of 1 million new 20-year-old men would be eligible for the draft each year.

CHAPTER THREE

To Go "Soldiering": Managing the Force Without a Draft

The previous chapter examined the history of the draft to try to understand when and under what conditions conscription has been used effectively to raise the manpower needed by the Army during wartime. Its main conclusion is that overwhelming support and equal sacrifice are the dual conditions needed before the American people will accept a draft. In this chapter, the report looks at how a volunteer force can be maintained, even during periods of conflict. Basically, the administration can initiate efforts to (1) increase the supply of volunteers to either enlist or reenlist into the armed forces, (2) reduce the demand for manpower by restructuring the current force, or (3) try to ameliorate the most negative aspects of deployment and family separation that result in military personnel and their families making the decision to leave the military.

Increasing the Supply of Volunteers

The most important difference between a conscripted force and a volunteer force is that the former is compelled to serve under penalty of law but the latter elects to serve without compulsion. While some may deride such incentives,[1] history has shown that volunteers increasingly

[1] According to military sociologist Charles Moskos, "extrinsic rewards, . . . can weaken intrinsic motivation" (Moskos and Wood, "Introduction—The Military: More Than Just a Job?" in Moskos and Wood, 1988), and the people we need in the military are not motivated by higher pay. The logical conclusion is that we cannot trust the soldier that we pay well because we can never be sure that he or she is not just serving for the money. If that forces

respond to bonuses and pay, with higher levels of compensation resulting in a greater number of volunteers.

From the Revolution to the Civil War

The uses of "bounties," or what today are called bonuses, to encourage soldiers to both enlist and reenlist is as old as the Army itself. While such "encouragements"[2] have been used during both periods of peace and of war, it is generally true that during difficult periods the "price" of volunteerism goes up. On January 19, 1776, General George Washington wrote to the Continental Congress urging them to "give a bounty of six dollars and two thirds of a dollar to every able bodied effective man, properly clothed for the service, and having a good fire lock, with a bayonet" (as quoted in ASD[M&RA], 1967a, p. I.1). This first enlistment bonus eventually grew to $200 by the end of the war (Kreidberg and Henry, 1955, p. 14). Within weeks, on February 9, 1776, faced with the prospect of needing troops for another year, Washington recommended to Congress that "they would save money and have infinitely better troops if they were, even at the bounty of twenty, thirty or more dollars, to engage the men already enlisted" (ASD[M&RA], 1967a, p. I.2).

The importance of bounties during the Revolution is well illustrated by the events of late 1776 and early 1777 surrounding the battle of Trenton. The battle of Trenton was one of the turning points of

---

a dedicated soldier to leave because he cannot support the quality of life he wants for his family, then Moskos's logic would hold that is all right because he has shown that he really did not have the calling in the first place (ASD[M&RA], 1967a). Empirical results, however, do not support Moskos's contention. A recent meta-analysis of the occupational commitment literature shows that "the demographic variable most strongly related to occupational commitment was income," possibly because "higher income increases commitment to organizations by enhancing one's self-esteem (Lee, Carswell, and Allen, 2000). Low wages are not a reflection of dedicated service but rather of the value that the American people put on military service. Many in Congress who voted for the large pay raises in the early 1970s, did so not because they favored the all-volunteer force but because they valued military service and thought that it was unconscionable for military wages to be below comparable civilian wages.

[2] George Washington asked Congress to approve bounties "for further encouraging the men more cheerfully to enter" (ASD[M&RA], 1967a).

the Revolutionary War, not least for the dire situation in which the Continental Army found itself at the end of 1776 after a series of disastrous defeats in New York earlier in the year. The spirit of the times was captured in Thomas Paine's second pamphlet, *The American Crisis*, with these words:

> These are the times that try men's souls. The summer soldier and the sunshine patriot will, in this crisis, shrink from the service of their country; but he that stands by it now, deserves the love and thanks of man and woman. (http://libertyonline.hypermall.com/Paine/Crisis/Crisis-1.html)

Indeed, there were many "summer soldiers" in Washington's army that Christmas. With the British and Hessian forces fragmented and settling into winter quarters along the northern side of the Delaware River, and with the prospect that a large portion of the army would fade away as New Year's and the end of enlistments rapidly approached, Washington decided to act. Breaking his army into three parts, Washington took a third of his force and forded the Delaware River north of the garrison town of Trenton, New Jersey. As the weather turned bad, his was the only part of the army able to cross in the early hours of Christmas morning, 1776. The bad weather, however, afforded him a degree of cover. His army took the Hessian garrison by surprise and, after a short fight, captured it and took many prisoners with minimal losses to his forces. After crossing back across the Delaware, Washington and his generals decided it would be best to remain on the offensive, but they realized that the end of the terms of enlistment were upon them. Historian David Hackett Fischer, in *Washington's Crossing*, tells us how Washington got the majority of those whose time was up to remain with the army for six more weeks of campaigning—campaigning that saw American victories in the second battle of Trenton and at Princeton:

> If Washington hoped to remain in the field, he had to persuade some of his veterans to stay with him. Finally, a solution was found by entrepreneurial officers of the Pennsylvania Associators,

> mostly Philadelphia merchants. They offered a bounty of ten dollars to men who agreed to turn out for a few more weeks of winter soldiering. . . .
>
> Washington was delighted by the result but appalled by the cost, . . . [but] agreed to try the same appeal with the Continentals in Greene's and Sullivan's division. . . . [He] spoke to the men and appealed to their conscience and honor. At the same time, he also addressed their material interest. Like General Mifflin he authorized a bounty of ten dollars in hard coin to every Continental soldier who agreed to stay, and he ordered the commanders at Morristown to do the same. It worked. (Fischer, 2004, pp. 271–273)

During the War of 1812, both the federal government and the states competed with bounties to encourage enlistments. Even when the federal enlistment bonus rose to $124—it had started at $8 at the beginning of the war—the states, under pressure to meet their obligation to raise their militias, outbid the federal government. After the war, with the principle of paying bounties for enlistments and reenlistments well established, Congress in 1838 authorized the first reenlistment bounty tied to grade and regular pay (e.g., three months' pay for a reenlistment of five years) (ASD[M&RA], 1967a, p. I.4). However, it was not just the American government that had to pay for volunteers; Great Britain, of course, needed to offer incentives as well.

The British Army of the 19th Century

In 1859, *Chambers's Journal* asked the proverbial question, why do men go "soldiering" when "small is the pay compared with the sufferings often endured?" (*Chambers's Journal* editorial staff, 1859b). In Europe at the time, the draft was taking upwards of 12 soldiers for every 1,000 inhabitants (and 14 in Russia). In Britain, with a volunteer force, the rate was eight per 1,000 inhabitants. (The American voluntary force today—active duty and Selected Reserve—draws about the same: eight per 1,000 inhabitants.)

Although wages and living conditions in the British military were below the standards for common British laborers, to encourage volunteers, Britain paid its soldiers considerably more than did other

European countries—37 percent more than in Belgium, 100 percent more than in France, and 278 percent more than in Russia (*Chambers's Journal* editorial staff, 1859b)—and provided them with better food rations, which included meat (*Chambers's Journal* editorial staff, 1859a). By the time of the Crimean War, British soldiers were enlisted—"engaged"—for ten years and received £3 as "bounty money" upon enlistment (*Chambers's Journal* editorial staff, 1859a). Given that "the poor and ignorant enter the ranks because the advantages are only sufficient to attract members of their class" (*Chambers's Journal* editorial staff, 1859b), the government also provided "trained army-schoolmasters . . . for each garrison and regiment" (*Chambers's Journal* editorial staff, 1859a). Soldiers and their children could attend for a nominal fee. Soldiers also received "good-conduct pay" for having many years of good service, as well as "beer money" and "fatigue pay" when they worked on public projects. They got a small "out-pension" when they retired, or "may have become weak and ailing after a moderate time, or may have been wounded in action" (*Chambers's Journal* editorial staff, 1859a).

Civil War Volunteers

From the outset of the Civil War in 1861 until the spring of 1863, the Union depended on militia and volunteers to fill the ranks, with the federal government paying a "bounty" of $100 to anyone who volunteered. By the spring of 1863, it was clear that the traditional system was producing unacceptable results. With industry booming in the Northeast, it was only the rural Middle West states that were meeting their quotas. Congress passed the first federal draft. The draft, however, produced less than expected. Of the first 300,000 called, more than 10 percent did not report; physical disabilities and dependency took 68 percent of the remainder; and of the 88,000 remaining, 52,000 bought their way out and 26,000 hired substitutes. Fewer than 10,000 were actually drafted into service (Chambers, 1987, p. 57).[3] With the need for manpower increasing and the three-year enlistment contracts of volunteers about to expire, on December 24, 1863 (Cutler,

---

[3] The actual figures are presented in Kreidberg and Henry (1955).

1922, p. 64), Congress raised the federal bounty to $300, which could be used for both enlistments and reenlistments. By November 1864, with states and localities competing with each other to attract volunteers who might help them meet their quotas without resorting to a draft, Congress raised the bounty to $600 (Cutler, 1922, p. 64)—six times what it was when the war started. The Civil War is known for the first federal draft; however, the vast majority—98 percent—of the 2.1 million men who fought under the Union banner were volunteers (Cutler, 1922, p. 80). Compensation, in the form of enlistment and reenlistment bounties, played an important part in the success of the Civil War volunteer system. It would be more than 100 years before America would use an enlistment bonus to raise its Army and maintain its Navy.[4]

Between the Civil War and the All-Volunteer Force

Regular pay and allowances were the norm for the American military from the end of the Civil War until the advent of the all-volunteer force, with a limited number of reenlistment bonuses and special incentive pays, including combat or hostile fire pay. Table 3.1 shows the various bonuses that were in effect in 2005.

Reenlistment Bonuses. Enlistment bonuses were not used between the Civil War and the transition to the all-volunteer force in 1971. Reenlistment bonuses, however, were justified to "protect an investment already made" and showed a willingness to put "a small portion of the high replacement training costs into the pockets of the already trained and experienced individuals, who are not now reenlisting" (statement by John A. Hannah, Assistant Secretary of Defense [Manpower and Personnel], June 10, 1954, as quoted in ASD[M&RA], 1967a, p. I.11).

Following the Civil War, it was not until 1908 that a reenlistment bonus was authorized, providing three months' pay. In 1912, it was expanded to four months' pay for a four-year reenlistment. During

---

[4] There was no enlistment bonus from the end of the Civil War until the passage of Public Law 92-129 in 1971, "when it became apparent that with the absence of the draft the supply of volunteers might not satisfy the requirements for new accessions resulting in a 'manpower gap' with the military services" (ASD[M&RA], 1967a).

Table 3.1
Comparison of Active Duty and Reserve Duty Bonuses

Active Duty Bonuses			Reserve Duty Bonuses			Consolidation
Authority	Service Requirements	Amount	Authority	Service Requirements	Amount	Significant Changes
Officer Accession Bonus	Accepts a commission and serves on active duty in a critical skill	Maximum of $60,000	None			**House Bill Section 618:** Authorizes a bonus up to $60,000 for accepting a commission or for an active duty officer affiliating with a reserve component
Enlistment Bonus	2-year minimum service obligation	$20,000 maximum	Non-prior-service enlistment bonus—Selected Reserve	None specified	$8,000 maximum	**House Bill Section 615:** Authorizes a bonus up to $20,000 for enlisting in a reserve component
			Prior-service enlistment bonus—Selected Reserve	3- or 6-year obligation	$8,000 for 6 years $4,000 for 3 years + $3,500 for second 3 years	
			Non-prior-service enlistment bonus—IRR	Authority expired on September 30, 1992, and has not been renewed	$1,000 maximum	
Reenlistment Bonus	3-year service	The lesser of $60,000 or 15 times the member's monthly basic pay times the service committment	Reenlistment bonus—Selected Reserve	3- or 6-year obligation	$5,000 for 6 years $2,500 for 3 years + $2,000 for a second 3 years	**House Bill Section 615:** Authorizes an enlisted member a bonus up to $60,000 for reenlisting in a reserve component
			Enlistment, reenlistment, or extension bonus—IRR	3- or 6-year IRR service obligation	$1,500 for 6-year agreement $750 for 3-year agreement	

Table 3.1—Continued

Active Duty Bonuses			Reserve Duty Bonuses			Consolidation
Authority	Service Requirements	Amount	Authority	Service Requirements	Amount	Significant Changes
Conversion Bonus to Critical Military Occupational Specialty	Minimum 3-year service commitment	Maximum $4,000	None			**House Bill Section 619:** Authorizes a bonus up to $4,000 for a reserve component member who agrees to convert a critically short military occupation
Criticial Skills Retention Bonus	Serve on active duty for at least one year	$200,000 maximum over the member's career	Affiliation bonus—Selected Reserve	Agrees to service in the Selected Reserve for the remainder of the member's military service obligation	$50 times the number of months remaining on the member's military service obligation	**House Bill Section 617:** Authorizes reserve officers and enlisted personnel a bonus of up to $200,000 over a career for agreeing to serve in a reserve component
Assignment Incentive Pay	Agree to serve on active duty in an assignment	$1,500 per month for each month in the assignment				**Senate Bill Section 617:** Assignment Bonus—establish a bonus for active duty and IRR members who agree to join a Selected Reserve unit (parameters similar to Assignment Incentive Pay)

SOURCE: William Carr, interview with author, February 11, 2005.

NOTE: IRR = Individual Ready Reserve.

World War I, and as a direct result of dissatisfaction with the Civil War draft, Congress provided that "no bounty shall be paid to induce any person to enlist in the military service of the United States" (as quoted in ASD[M&RA], 1967a, p. I.8). After World War I, in 1922, Congress authorized an "enlistment allowance" of $50 multiplied by the number of years served in the enlistment period for anyone in the first three grades. A multiple of $25 was authorized for all other grades. Despite the name change, the concept was the same as the reenlistment bonus. The enlistment allowance remained until 1933—when it was canceled as an economy move during the Depression—and was reinstated on June 16, 1942. After World War II, the Career Compensation Act of 1949 used for the first time the term "reenlistment bonus." The act provided variable lump-sum payments for reenlistments of from two to six years, regardless of grade (ASD[M&RA], 1967a, p. I-10). In 1954, the Reenlistment Bonus Act returned the grade of a service member into the calculation of the reenlistment bonus, basing the bonus on the member's monthly base pay multiplied by the number of years reenlisted. To further encourage reenlistments, the Cordiner Committee, appointed to study the problem of low retention, concluded that "[s]till greater monetary incentive is required to strengthen the motivation features of enlisted compensation" (as quoted in ASD[M&RA], 1967a, p. I.12). Setting a theme that anticipated the all-volunteer force, the committee recommended that pay provide a "direct and selective monetary inducement to improve personnel retention and job motivation" (as quoted in ASD[M&RA], 1967a, p. I.12). The resulting higher "proficiency pay" was tied to shortages in a military specialty, special duty assignments, and superior performance. In 1965, the Variable Reenlistment Bonus was introduced to further encourage reenlistments in "specific critical skills."

Pay and Enlistment Bonuses. The Pentagon's Second Quadrennial Review of Military Compensation in 1971 summed up the prevailing views on military compensation before the advent of the all-volunteer force:

> Historically [or at least since the beginning of World War II], the draft has acted to provide the majority of first termers. [The]

> pay philosophy applied in the past has included the ideal of the citizen's obligation to serve a minimum period in uniform. This philosophy accounts for the omission of the first termer from several periodic pay increases granted in the past to the remainder of the force. As a result, at the end of 1970 the Armed Forces found that many new entrants were paid below the Federal minimum wage. For the new entrant with a family, maintaining an adequate family life on military wages was very difficult. (ASD[M&RA], 1967b, p. 2)

However, the passage of Public Law 92-129 in September 1971 changed this situation. First-term pay was dramatically increased, and an enlistment bonus of $3,000 was authorized for the first time since the Civil War. Today, with the military engaged in Afghanistan and Iraq, enlistment bonuses can be as large as $20,000 and reenlistment bonuses as large as $60,000.[5] And sometimes, under special circumstances, reenlistment bonuses are tax-free.

Hostile Fire Pay and Hazardous Duty Pay. Providing extra pay for those engaged in combat dates only to World War II. (Table 3.2 shows the full range of "away pays," including hostile fire pay in effect in 2002.) The initial proposal for "fight pay" was to "provide special recognition for the infantryman who endures the greatest hardship and suffering in time of war" (ASD[M&RA], 1967c, p. I.2). When passed in 1945, the pay was retrospectively tied to the awarding of the Expert Infantry Badge and the Combat Infantry Badge. The extra pay continued to be paid to servicemen who held the badges until October 1949.

At the beginning of the Korean War, the Army wanted to give "all personnel who are engaged with the enemy" hazardous duty pay instead of having a separate hostile fire pay.[6] As finally enacted, however, the new "combat duty pay" remained separate. It differed from hazardous

---

[5] The *Baltimore Sun* reported that the Pentagon is offering as much as $150,000 for Green Berets and Navy SEALs to reenlist for six years. When comparing dollar amounts over time, however, one must be aware of the general changes in prices and wage levels. To be effective, bonuses must increase in nominal dollars just to retain real value (Bowman, 2005).

[6] The Army's proposal is discussed in ASD(M&RA) (1967c).

Table 3.2
"Away Pays" in Effect, 2002

Pay	Paid For	Amount	Varies With	Other Restrictions
Career Sea Pay	Assignment to ship	$50–520/month; average $200 for E-6	Pay grade and cumulative sea duty	Pay grade E-4 and above
Career Sea Pay Premium	Over 36 continuous months assigned to sea	$100/month	Fixed	Paid to E-4s and officers; E-5s to E-9s up to the fifth year of sea duty
Submarine Duty Pay	Operational sub duty for lower pay grade; sub-qualification for higher pay grades	$75–355/month; average $200 for E-6	Pay grade and years of submarine service	
Family Separation Allowance	Enforced family separations	$100/month; prorated daily	Fixed	Must have spouse and/or dependents and be away more than 30 days
Hostile Fire/ Imminent Danger Pay	Subjected to hostile fire, hostile mine, or threat thereof	$150/month	Fixed	IDP
Hardship Duty Pay—Mission	Designated hardship mission (e.g., prisoner of war remains recovery)	$150/month	Fixed	
Hardship Duty Pay—Location	Living conditions far below those in the U.S.	$50–150/month	Severity of hardships	OCONUS locations TAD or PDC in excess of 30 days
Overseas Tour Extension Incentive Pay	Extending OCONUS tour at least one year	$80/month or extra leave	Fixed	Paid to specific occupational specialties
Combat Zone Tax Exclusion	Serving in designated combat zone	Taxes on basic and some special pays	Income level	Officer income exclusions have upper limits
High Employment per Diem	Days deployed in excess of 400/730	$100/day	Fixed	

SOURCE: Office of the Under Secretary of Defense (Personnel and Readiness) (2002), Vol. 1, p. 112.

NOTE: IDP = Imminent Danger Pay; OCONUS = outside the continental United States; TAD = temporary additional duty; PDC = permanent duty change.

duty and other incentive pays in that it was only for a period of hostilities and was the same for all pay grades. It may best be described as "gratitude" pay, providing "a critical degree of recognition of the rigors of war endured by those in combat" (ASD[M&RA], 1967c, p. I.15).

Recognition was again afforded for combat service at the beginning of the war in Vietnam. Noting the engagement of American forces without the formality of a declaration of war, Congress authorized hostile fire pay "except in time of war declared by the Congress" (Ogloblin, 1996). In 1983, the notion of a peacetime conflict saw Congress extending the additional pay to service members serving in foreign areas that were subject to "the threat of physical harm or imminent danger on the basis of civil insurrection, civil war, terrorism, or wartime conditions" (Ogloblin, 1996, p. IID.1.a.[1]). The pay was renamed "hostile fire or imminent danger pay." During CY 1992, as a result of Operation Desert Storm, more than 327,000 troops received hostile fire or imminent danger pay. For the rest of the decade, reflecting frequent deployments to the Gulf and the Balkans, upwards of 60,000 troops received hostile fire or imminent danger pay. As much as anything else, the numbers of troops receiving this pay served as a barometer of the level of engagement and the stress the troops were under, even before the current situation in Afghanistan and Iraq.

Compensating a Force Already Under Stress: From Bosnia to Iraq. Since the end of the Cold War, the American military has constantly been engaged in conflicts resulting in what Secretary Rumsfeld describes as being "under a great deal of stress" (Rumsfeld, 2004b). Both the Clinton and Bush administrations have substantially increased pay to compensate the force. To date, these pay changes have sustained the active-duty force, and while recruiting has fallen somewhat short of requirements recently (in FY 2005), retention has remained high.[7] The

---

[7] On July 26, 2004, the American Services Press Service reported, "The overall [active duty] Army retention rate is more than 100 percent. This overall rate is broken into three categories—initial, mid-career and careerists. The first term re-enlistment rate is over 100 percent of goal. The careerists are at or over 100 percent also. The mid-career soldiers—those between six and 10 years of service—are experiencing a dip in re-enlistments (Garamone, 2004a). "The Army, Navy, Air Force and Marine Corps all made their active duty recruiting numbers for fiscal 2004. . . . The Army enlisted 77,587 soldiers through September, besting the year's

situation with the reserve forces is less sanguine, particularly recruiting for the Army National Guard.[8]

A Din of Complaints—Personnel Tempo (PERSTEMPO). Since the end of the Cold War, a central issue has been, "What to do about the unexpected increase in the tempo of operations?" Through the late 1990s, the deployment of troops to Bosnia, continued air patrols over Iraq, and naval patrols in the Persian Gulf were generating a din of complaints. As so often happens, it was the service members in the field and their complaints to Congress that first alerted personnel managers to the problem. In response to a request from the Chairman of the House Subcommittee on Military Readiness and Military Personnel, the General Accounting Office (GAO; now the Government Accountability Office) showed how sharply deployments had increased from 1990 to 1995, especially for the Army and the Air Force. The GAO concluded that "[t]he time military personnel are spending away from home on deployments—PERSTEMPO—has increased and is stressing portions of the military community and adversely affecting readiness" (GAO, 1996, p. 1).

In fact, as early as 1995, the GAO was warning of a force structure/mission requirement mismatch. In what became known as a high-demand and low-density problem, it noted: "Peace operations heavily stress some U.S. military capabilities. . . . Repeated use of these forces, of which there are relatively few in the active force, has resulted

---

goal by 587 soldiers. Through Sept. 29, the Navy reported that it enlisted 39,874 sailors, bettering its goal by 254 sailors. The Air Force . . . enlisted 34,362 service members for the year, topping its recruiting goal by 282 people. The Marine Corps . . . enlisted 36,794 service members for fiscal 2004, which topped its goal by 21 enlistees. The Army [had] . . . fewer enlistees enrolled in its delayed-entry program for fiscal 2005. . . . [The Army] recently fielded more recruiters and made more aggressive use of bonuses in order to attract and sign up more recruits" (Garamone, 2004a).

[8] In the summer of 2004, the Sergeant Major of the Army told the American Forces Press Service, "For the Army Reserve the picture is also fairly clear. The component is at 98.7 percent, well on the glide path for accessions. However, for the Army National Guard accessions are at 87.2 percent. What we think is that you have a lot of active duty soldiers who are re-enlisting to stay in the Army. . . . Those that are getting out are not necessarily going in to the National Guard or (Army) Reserve. . . . On retention, the National Guard was at 118 percent, so the two kind of balance each other" (Garamone, 2004a).

in some units and personnel deploying more than once to an operation or to consecutive operations, increasing the tempo of operations" (GAO, 1995, p. 4).

To reduce deployments and reduce the stress on service members and their families, Congress mandated "burdensome tempo pay," or "high deployment per diem." Under the congressional plan, any service member deployed more than 250 days in the previous year would receive $100 per day for each additional deployment day.[9] The plan was to go into effect in 2002 but was suspended on October 8, 2001, a day after the beginning of Operation Enduring Freedom. In 2002, Under Secretary of Defense David S.C. Chu wrote the Chairman of the Armed Services Committee, suggesting a number of changes to "streamline current management thresholds and required actions . . . [and] improve [the] structure, levels and flexibility of compensation to members" (Chu, 2002, p. 5). The National Defense Authorization Act of 2004 incorporated a new set of changes recommended by DoD. The new law "authorizes payment of a monthly high-deployment allowance of up to $1,000, instead of the $100 high-tempo per diem allowance . . . for service members each month during which the member is deployed for 191 or more consecutive days or for 401 days out of the preceding 730 days" (Committee of Conference, 2003, p. 694).

In 2004, with the average number of "away from home" days almost having doubled from FY 2001 to FY 2004—as shown in Table 3.3—DoD proposed the "Triple Backstop" system, which has three components. First, members who are sent to less-desirable locations are compensated using high-deployment pay (location); the rates are established by country. Second, high-deployment pay (tempo) will compensate those who are deployed excessively—"too long and/or too frequently"—with the definition and amount of compensation left to each service to determine. Third, the existing Selective Reenlistment Bonus would be made part of this program. The goals of all these

[9] U.S. Code, Title 36, Section 435. In 2001, Congress changed the threshold for the new pay from 250 days out of 365 days to 400 days out of 730 days.

Table 3.3
Service Size and Deployment Summary, FY 2001 to FY 2004

		FY 99	FY 00	FY 01				FY 02				FY 03			
					Deployed				Deployed				Deployed		
Component		End Strength	End Strength	End Strength (ES)	Members Deployed	Avg Days/ Deployed Members	Avg Days/ES	End Strength (ES)	Members Deployed	Avg Days/ Deployed Members	Avg Days/ ES	End Strength (ES)	Members Deployed	Avg Days/ Deployed Members	Avg Days/ES
Army	Active	479,426	482,170	475,072	255,853	51.3	27.6	477,914	202,969	81.1	34.4	488,640	226,274	160.8	74.4
	Reserve	391,409	369,215	362,295	74,214	17.2	3.5	337,015	81,135	34.8	8.4	337,015	80,586	65.9	15.7
	Guard	362,059	357,257	355,351	189,578	17.5	9.3	354,293	199,065	49.3	27.7	350,568	219,230	108.9	68.1
	Total	1,232,894	1,208,642	1,192,718	519,645	34.1	14.9	1,169,222	483,169	60.2	24.9	1,176,223	526,090	124.6	55.7
Navy	Active	373,046	373,193	366,990	183,340	118.9	59.4	376,781	190,915	121.3	61.4	377,881	177,726	126.7	59.6
	Reserve	202,411	191,293	172,681	61,305	87.3	31.0	154,525	51,798	98.7	33.1	154,525	44,159	105.3	30.1
	Total	575,457	564,486	539,671	244,645	111.0	50.3	531,306	242,713	116.4	53.2	532,406	221,885	122.5	51.0
Marine Corps	Active	172,641	173,321	171,688	96,756	67.6	38.1	171,142	96,672	84.4	47.7	176,087	109,294	126.7	78.6
	Reserve	99,388	100,750	98,109	9,376	15.6	1.5	96,570	15,411	73.1	11.7	96,570	25,989	147.5	39.7
	Total	272,029	274,071	269,797	106,132	63.0	24.8	267,712	112,083	82.9	34.7	272,657	135,283	130.7	64.8
Air Force	Active	360,590	355,654	348,821	190,178	43.7	23.8	357,392	190,666	55.9	29.8	366,278	206,626	69.8	39.4
	Reserve	143,172	139,073	191,308	46,775	25.4	6.2	114,433	38,905	46.8	15.9	114,433	35,258	49.3	15.2
	Guard	105,715	106,365	121,891	55,833	21.3	9.8	111,242	40,717	42.3	15.5	109,457	39,722	45.0	16.3
	Total	609,477	601,092	622,020	292,786	36.5	16.2	583,067	270,288	52.6	24.4	590,168	281,606	63.7	30.4
DoD	Total	2,689,857	2,648,291	2,664,206	1,163,208	53.5	23.4	2,561,307	1,108,253	73.0	31.7	2,571,454	1,164,864	110.2	49.9

SOURCE: Rumsfeld (2004a). Data from the Defense Manpower Data Center.

NOTE: Prior to 2001, the services did not consistently track deployed data.

efforts are to "adequately" compensate members who are subject to long and/or frequent deployments and to positively affect the decision to reenlist (Carr, 2004). The problem, however, of keeping track of the deployment days for individual members remains. In 2002, Under Secretary Chu described it as "a new burdensome requirement on virtually every unit/organization, from the unit level to the Department level" (Chu, 2002). For the Army, short of the return of the venerable company clerk, it is unlikely that accurate records can be maintained, resulting in an administrative morass, as a soldier's claim for "time away from home" will be difficult to corroborate through existing information systems.

Fixing the Basic Pay Table. Throughout the 1980s, the report of high levels of career retention by the Secretary of Defense was one indication of the improving fortunes of the all-volunteer force. By 1988, research at RAND was showing that only 3 to 4 percent of enlisted personnel from the draft era cohort (FY 1967–FY 1970) reached retirement eligibility compared with projections of about 18 percent for the more recent FY 1987 cohort. When finally tested in Operation Desert Storm in 1991, the career-oriented all-volunteer force performed extremely well.[10] By the summer of 1998, however, Secretary of Defense William Cohen knew something was wrong, especially with the career force.

On September 15, 1998, Secretary Cohen, the Joint Chiefs of Staff, and the heads of the unified commands met with President Clinton to discuss what Chairman of the Joint Chiefs of Staff Gen Henry Shelton called the "nosedive" in readiness. Although they discussed the need to balance readiness and procurement, Cohen's spokesman, Ken Bacon, emphasized that the challenge was not to overlook "retention issues like military pay and retirement benefits" (Gillert, 1998).

The FY 2000 budget contained revised pay tables. These pay changes substantially reformed the basic pay table by giving larger pay

---

[10] Nick Timenes, the Principal Director in the Office of Military Manpower and Personnel Policy, summed up the situation this way: "The all-volunteer force worked. It took a generation to get here, but in Desert Storm the enlisted force exhibited unprecedented skill, commitment, maturity, and professionalism" (Timenes, 1991).

raises to the career force and, by at least one account, are an important reason military retention has remained high in the face of extended deployments and combat casualties in the war in Iraq.[11] Targeted pay raises for midlevel officers and noncommissioned officers (NCOs) continued until 2003 (Gilmore, 2004). In addition, Congress was concerned about the deleterious effects of increases in the tempo of operations service members faced and attempted to initiate PERSTEMPO pay, not only to compensate service members for extra work, but also to penalize the services for causing the extra work in the first place.

Volunteerism in Meeting Military Commitments Around the Globe. In his letter to the Armed Services Committee in 2002, Under Secretary Chu expressed his hope that in the future DoD might "harness volunteerism to the task of meeting military commitments around the globe" (Chu, 2002, p. 7). The Navy "tested" the idea that the burden of being assigned to a less desirable place might be made less onerous if the sailor volunteered for the assignment rather then being forced to go. In 2003, the Navy introduced the Assignment Incentive Pay (AIP) program as a "pilot program." In a message to the fleet, the Chief of Naval Operations (CNO) wrote:

> AIP will enhance combat readiness by efficiently distributing sailors where they are most needed. This pilot program will offer sailors up to $450.00 per month for assignments [to certain locations]. . . . AIP will employ a market-based approach, allowing sailors to set the "price" for a particular assignment (below the Navy established maximum). [The] Navy can therefore meet sailors' expectations of a fair incentive for the assignment and improve fleet manning. (CNO, 2003)

In effect, the Navy will be running a reverse silent auction. A sailor making a relatively low bid will be taken before a sailor making a relatively high bid. The sailor sets his price. The CNO reminded sailors, "Other sailors may also be bidding for the same assignment." In practice, two sailors working side by side would surely have made dif-

---

[11] A view expressed by William Carr, Acting Deputy Under Secretary of Defense for Military Personnel Policy, interview with author, September 22, 2004.

ferent bids and, as a result, would be getting different AIP payments. Two months after the program started, the Navy reported, "Response has been excellent" (Chief of Naval Personnel Public Affairs, 2003b); in December 2003, the Navy expanded the program to include service aboard repair ships (tenders) based in Sardinia and Guam (Chief of Naval Personnel Public Affairs, 2003a).

While the Navy emphasizes the benefits of "giving sailors the power of choice" (Chief of Naval Personnel Public Affairs, 2003a), some outside the Navy object to the extension of volunteerism in this way. LTG Ron Helmly, Chief of the Army Reserves, in his letter to the Chief of Staff of the Army, complained that "to use other than involuntary mobilization authorities places the burden of responsibility on the Soldiers' back instead of the Army's back" (Helmly, 2004). Helmly further expressed his concern that "to incentivize 'volunteers' for remobilization by paying them $1,000 per month" would cause "potential 'sociological' damage" to the force. He argued, "The use of pay to induce 'volunteerism' will cause the expectation of always receiving such financial incentives in future conflicts" (Helmly, 2004).

General Helmly's concerns notwithstanding, the Army extended AIP to assignments to Korea for between $300 and $400 a month. In the first three months of the program, more than 7,500 soldiers signed up for additional duty in Korea. "I was pleasantly surprised at how well the Assignment Incentive Pay (AIP) has worked out there," the Sergeant Major of the Army wrote in his leader's notebook on the Army's Web site.[12] In April 2004, the Air Force started its own AIP program, also for Korea.[13]

Using Incentives to Sustain the Force During Periods of Conflict. The notion that an all-volunteer force might be sustained during peri-

---

[12] He went on to say, "I reminded the Soldiers that for each one of them who had agreed to stay in Korea the extra year or two, they had saved the Army two PCSs, one to Korea and one back from Korea. I received no complaints about the AIP and found more than 9,000 Soldiers serving there have opted to participate in the program. I think the Soldiers in Korea, much like the Soldiers in the 3rd ID [Infantry Division] are doing an incredible job and I commend them as well as their leadership for what they do every day" (Preston, 2004).

[13] William Carr, Acting Deputy Under Secretary of Defense for Military Personnel Policy, interview with author, February 11, 2005.

ods of conflict through the use of incentives was new and untried before the current war in Iraq. Crawford Greenewalt, a member of the Gates Commission, wrote to Thomas Gates in 1969 as the commission was completing its work, "While there is a reasonable possibility that a peacetime armed force could be entirely voluntary, I am certain that an armed force involved in a major conflict could *not* be voluntary" (Greenewalt, 1969, emphasis in the original).

On September 3, 2003, the head of the Congressional Budget Office (CBO) told senior members of Congress:

> The active Army would be unable to sustain an occupation force of the present size—180,000, about 150,000 deployed in Iraq itself and the rest supporting the occupation from neighboring countries—beyond about March 2004 if it chose not to keep individual units deployed to Iraq for longer than one year without relief. (Holtz-Eakin, 2003)

To sustain such deployments, the CBO concluded that

> DoD could seek the authority to use temporary financial incentives to increase the number of personnel that could be sent to Iraq. Such incentives could encourage current selected-reserve and active-duty personnel to voluntarily accept higher deployment tempo and induce new categories of reserve personnel or prior service members to volunteer for deployment. (Holtz-Eakin, 2003, p. 26)

The CBO admitted, however, that "DoD does not have experience using bonuses to encourage military personnel to deploy voluntarily to a hostile area . . . [and] thus the effects of offering such financial incentives are unknown" (Holtz-Eakin, 2003, p. 26).

New financial incentives have been developed for both recruiting and retaining the personnel needed today. The $420 billion National Defense Authorization Act of 2005 continued a full range of recruiting and retention bonuses, as well as extended health benefits for some reservists, and provided a new educational assistance program for the reserves tied to the Montgomery GI Bill. As the FY 2006 budget was submitted to Congress, Secretary Rumsfeld noted, "Military pay has increased about 25 percent, . . . [with the FY 2006 budget including] a 3.1 percent increase in base pay, plus bonuses, and recruiting and retention programs to ensure the Defense Department maintains its professional fighting force" (Rumsfeld, 2005).

A service member serving in Iraq is entitled to regular military compensation and a myriad of special pays and benefits. The National Defense Authorization Act for Fiscal Year 2005 provided a permanent increase in hostile fire/imminent danger pay of $225 per month and in family separation pay of $250 per month (Garamone, 2004b). In addition, a soldier gets $100 per month hardship duty pay (location) for serving in Afghanistan or Iraq. Many soldiers prefer to reenlist during a deployment to a combat zone because the pay received there—including reenlistment bonuses, which can run up to $15,000—is tax free.[14] The combat zone tax benefit exempts all income earned in the combat zone by enlisted personnel and warrant officers from federal income tax. The income of officers subject to the exemption is limited. States generally provide a similar tax exemption. Those serving more than three months in Afghanistan or Iraq are also eligible to deposit up to $10,000 in a special Savings Deposit Program account that pays 10-percent interest, a "significantly increased amount" compared with their Thrift Savings Plan account. Also, part of a soldier's federal student loans can be "forgiven" for service in a combat zone. Finally, for those soldiers involuntarily extended beyond the 12-month service tour and members of the reserve components who volunteer to remain on active duty beyond their cumulative 24-month mobilization duty to complete 12 months in country, DoD uses its AIP funds to provide "additional special compensation" of $1,000 per month. The military also provides special compensation to service members with special skills who volunteer to extend in Iraq and Afghanistan past 12 months of between $300 and

[14] Some reenlistment bonuses are substantially higher. The Navy is offering $60,000 bonuses "to retain high quality . . . SEAL personnel . . . as part of an overarching incentives plan to compensate our special operations forces for their contribution and sacrifice while in support of the global war on terrorism" (CNO, 2005). When comparing dollar amounts over time, however, one must be aware of the general changes in prices and wage levels. To be effective, bonuses must increase in nominal dollars just to retain real value.

$1,000 per month, depending on the length of the extension.[15] Table 3.4 compares the additional compensation for a typical E-4 (who is married with one child) serving in these areas with that of his stateside counterpart. It does not include the "additional special compensation" discussed above.

The Rising Cost of Manpower. Although using financial incentives to attract and retain military personnel seems to have been generally-successful in allowing DoD to maintain the size of the active military,[16]

Table 3.4
Comparison of Pay for Soldiers

Monthly (E-4, 4 years of service, married, 1 child)	CONUS Station (in garrison, with family)	Iraq or Afghanistan (1-year Temporary Duty)
Basic Pay[a]	$2,596	$2,596
Basic Allowance for Housing (BAH)[b]	$1,106	$1,106
Basic Allowance for Subsistence (BAS)	$254	$254
Family Separation Allowance (FSA)	$0	$250
Temporary Duty—Per Diem (Incidental Expense)	$0	$105
Hardship Duty Pay—Location (HDP-L)	$0	$100
Imminent Danger Pay (IDP)	$0	$225
Combat Zone Tax[c]	$0	$90
Total	$3,103	$4,726
Difference (from CONUS station)		$770

a Based on January 1, 2004, pay table.

b Assumes average BAH for all E-4s with dependents. Actual BAH rate for an individual member would be determined based on geographical location.

c Assumes no spousal income. Relects 2004 tax rates.

---

[15] William Carr, Acting Deputy Under Secretary of Defense for Military Personnel Policy, interview with author, February 11, 2005.

[16] See Assistant Secretary of Defense (Public Affairs) (2005).

it comes at a substantial cost. Cindy Williams of the Massachusetts Institute of Technology's Security Studies Program reports that "[t]he total cost of military pay and benefits increased by nearly 29 percent between 2000 and 2004—three and one-half times the rate of consumer inflation and about twice the rate of wage inflation in the private sector" (Williams, 2005). The issue was so much of a concern that in 2004 the Pentagon sponsored a conference at the Institute for Defense Analyses to address the subject. In his opening remarks, Ken Krieg, Director of the Office of Program Analysis and Evaluation, noted his concern that

> the fully loaded cost of manpower is growing rapidly. It is the only consistently growing part of the defense budget, yet managers throughout the Department of Defense act as though uniform manpower is free . . . —that the price of the good that is growing rapidly is disassociated from how people deal with things—is a huge issue. (Horowitz and Bandeh-Ahmadi, 2004, p. 3)

Under Secretary Chu addressed Krieg's concerns and told the conference:

> I do think it's important to keep in mind that the military compensation system, whatever its idiosyncrasies, does work reasonably well in producing the results that we want. . . . It's critical to keep in mind the compensation system is not an end of itself. . . . The system is, after all, an instrument to reach the results we want, which is to supply young Americans who are willing to take on some of the most difficult and demanding tasks that society might ask them to do. It's not the only reason they serve, but it's an important element of their decision to serve, and it's certainly important in their family's decision to support such service. . . . [All] too often these debates turn into the question of "how can I save money," not "how can I produce the intended results." . . . I do think we need to move in the resource community away from an undue focus on how much things are going to cost. Cost is important and we want to be efficient, but it is critical to start with what . . . [we want] to achieve. (Horowitz and Bandeh-Ahmadi, 2004)

Reducing Demand by Transforming the Force

In 2004, Secretary Rumsfeld told the Chairman of the House Armed Services Committee that the force was "stressed" because it was "not properly aligned or organized for the post–Cold War era" (Rumsfeld, 2004c). As he saw it:

> Too many skills we need are heavily concentrated in the reserve components. Too many of our active forces are organized in large, heavy divisions that are not readily deployable. Too many military personnel—tens of thousands—are performing tasks that could and should be performed by civilians. (Rumsfeld, 2004c)

The solution to the problem, he thought, was

1. To increase the size of the Army by 30,000 troops
2. To increase the number of deployable brigades from 33 to 43, with the goal of reducing the frequency of, and increasing the predictability of, deployments
3. To "rebalance" skills between the active and reserve components.

Failure to Restructure After the Cold War

It is often said that generals plan to fight the last war. Certainly, Secretary Rumsfeld's observation coming almost a decade and a half after the fall of the Berlin Wall suggests the validity of that old adage. The numerous efforts[17] to "rethink" what America's military force structure should be after the end of the Cold War all had at their core a vision of conventional warfare—a vision that a small minority called a "mistake."

Even before the current situation in Iraq, some were arguing that the problems the United States had in Somalia in 1992 and 1993 were more indicative of future military operations in the post–Cold War era

---

[17] Initially, Secretary of Defense Richard Cheney and Chairman of the Joint Chiefs Colin Powell had the *Base Force*; later, the Clinton administration made some modifications with its *Bottom-Up Review* and *Quadrennial Defense Review*; and in the current administration, Secretary Rumsfeld has pushed *Transformation*.

than was Operation Desert Storm. Gen. Anthony Zinni, USMC (Ret.), the commander who oversaw the withdrawal of American troops from Mogadishu and later the commander in chief of the Central Command, told attendees of the Robert McCormick Tribune Foundation–Naval Institute Annual Seminar in 2000, "Operation Desert Storm, as far as I am concerned, was an aberration. . . . In still trying to fight our kind of war, be it Desert Storm or World War II, we ignore the real war fighting requirements of today" (Zinni, 2000).[18]

The failure in Somalia to accomplish U.S. objectives exposed deficiencies in U.S. military posture that would later strain the very fabric of the all-volunteer force. After 50 years of engagement with the Soviet Union, the United States had a force structure and a set of all-volunteer personnel policies designed to attract, retain, and motivate personnel to man a "conventional" force. If such a force proved less than adequate for a Somalia-type "operation other than war" (OOTW), no one should blame the all-volunteer force. In Somalia, the planning assumption that an OOTW engagement was the lesser but included case of conventional war proved wrong. Understanding the nature of future military engagements, defining the appropriate force structure for the future, and developing an appropriate set of personnel policies for an all-volunteer force should have been the legacy of Somalia, but it was not. The force the United States took to Afghanistan in 2002 and Iraq in 2003 was optimized for a "conventional war." The remarkable U.S. success during the first weeks of the two wars—both conventional wars seemed to validate all the United States had concluded about the American way of war and its volunteer force. It did not, however, fore-

---

[18] General Zinni was not the only one who thought more attention should have been paid to OOTW. For example, Allard (1995) suggested a number of lessons and provided some insights that, if adopted, might have better prepared us for the recent operations in Iraq. The Congressional Research Service reviewed the post–Cold War commitment to such operations in light of the events of September 11, 2001, also noted: "Technology advances made transforming U.S. forces even more combat effective against conventional forces, but could not yet substitute for all the manpower needs in the unconventional and asymmetric environments of 'stability' operations. In contrast, some charge that the Army, in particular, was resisting such 'constabulary' operations and therefore managed its personnel inefficiently" (Bruner, 2004).

tell the difficulties the United States later had in winning the peace or the strains this type of operation placed on the all-volunteer force. Citing joint doctrine (Clark, 1999), Ken Allard underscored this:

> Peacekeeping requires an adjustment of attitude and approach by the individual to a set of circumstances different from those normally found on the field of battle—an adjustment to suit the needs of peaceable intervention rather than of an enforcement action.
>
> In addition to the individual character traits discussed by that publication, the most important ones are probably good judgment and independent action.
>
> Enforcement actions require all these things in addition to the ability to transition rapidly to full-scale combat operations when required. MG Montgomery[19] has noted the need for more effective predeployment training standards, including the in-theater ROEs [rules of engagement], local culture, and weapons familiarization. . . .
>
> One final point: peace operations put a premium on certain specialists who should be identified early and placed near the front of any deployment—possibly on the first plane. They include: trained Joint Operations Planning and Execution System (JOPES) operators, contract specialists (especially those with experience in local procurement), logisticians, lawyers, medical specialists, . . . port transportation organizers, public affairs officers, military police, combat engineers, psychological operations specialists (PSYOPS), and civil affairs experts,[20] as well as special forces teams. Equally important are people with specific knowledge of the language and the country. . . . The use of Reserve Component personnel with special qualifications for service in Somalia also worked well—suggesting the importance of Reserve Component integration in the planning of future peace operations. (Allard, 1995)

---

[19] The Montgomery Report is a classified report by the Center for Army Lessons Learned at Fort Leavenworth, Kansas.

[20] Others have also commented on the importance of PSYOPS as a "force multiplier." For examples, see Hoffman (2004).

The War in Iraq

Early on the morning of March 20, 2003, American and coalition troops attacked Iraq. Baghdad fell on April 10, 2003. On May 1, 2003, on board the USS *Abraham Lincoln* off the coast of San Diego, President George W. Bush announced that major combat operations had ended (CNN.com, 2003). With the end of conventional combat operations, the Army and Marine Corps entered a period that some have called "nation-building." Rather than reducing troop levels in Iraq as planned, on May 4, 2004, Secretary Rumsfeld announced, "The overall U.S. troop strength in Iraq will be stabilized at approximately 138,000 as requested by the combatant commander." The announcement also confirmed that "various units from the National Guard and Reserve are in the deployment. . . . All Army National Guard and Reserve units being deployed will be given sufficient time to train in preparation for their service in Operation Iraqi Freedom" (Office of the Assistant Secretary of Defense [Public Affairs], 2004). The reserves would be deployed up to 12 months in Iraq, but the total time they would be away from home would "depend on training requirements and the requirements of the Central Command commander" (Office of the Assistant Secretary of Defense [Public Affairs], 2004).

Rebalancing the Force

Even before the beginning of Operation Iraqi Freedom, the need to rebalance the force was recognized in the December 2002 *Review of Reserve Component Contributions to National Defense* (Rumsfeld, 2003). The report highlighted a number of "indicators" that suggested something was wrong, and cited the following:

> Routine use of involuntary recall of the reserves; increased operational tempo in selected areas; anecdotal evidence that the ongoing partial mobilization may have a negative impact on reserve recruiting and retention in the future; the mismatch between the new defense strategy and current force structure; and the length of time it takes to adapt force-mix allocations in today's rapidly changing security environment. (Rumsfeld, 2003)

The report noted,

> Contingencies such as peacekeeping and humanitarian operations place a high demand on some capabilities—civil affairs, military police and security forces, public affairs units, air traffic control services, deployable air control squadrons, and the reserve intelligence community—that are low in density to overall available forces . . . are high in demand as the Department strives to meet global security requirements. (Rumsfeld, 2003)

Rumsfeld attributed this high demand/low density of some specialized forces to the assumptions that these "capabilities . . . [were thought to be needed] only in the later phases of a conflict under the two-major-theater-war strategy" (Rumsfeld, 2003).

In early 2004, the Deputy Assistant Secretary of Defense for Reserve Affairs reported on efforts to rebalance forces by moving people from low-demand positions to fill vacancies in high-demand positions. The theme of the program, he noted, was "to improve the responsiveness of the force and to help ease stress on units and individuals with skills in high demand" (Winkler, 2004, p. v). Over the previous three-year period, he reported, the services changed 50,000 military spaces—10,000 in FY 2003 and 20,000 each in the two subsequent years. The reported rebalancing was based on a December 2002 *Review of Reserve Component Contributions to National Defense* and the Secretary of Defense's directions of early July 2003, which instructed the services to "restructure active and reserve forces to reduce the need for involuntary mobilization, . . . establish a more rigorous process for reviewing joint requirements . . . provide timely notice of mobilization [and] make the mobilization and demobilization process more efficient" (Rumsfeld, 2003). Among other things, Secretary Rumsfeld also established the planning goals of using a guardsman or reservist "not more than one year every 6 years" (Rumsfeld, 2003).

While the rebalancing is being carried out by all the services, it is the Army that has drawn the most attention, given the situation in Iraq. To increase Army readiness without having to call up reserve forces with little or no warning, the "Rebalancing Forces" report noted: "The Army is converting 5,600 spaces of lower priority active structure to higher priority active structure. These conversions will add capabilities in chemical, military police, engineer (bridging and fire fighting units), medical, quartermaster (fuel, water, and mortuary affairs units), and transportation specialists" (Rumsfeld, 2003, p. 11). Recognizing that the "global security environment" was placing stress on certain career fields, in FY 2001 the Army reprogrammed 30,000 spaces,

> providing additional capabilities in the areas of civil affairs, psychological operations, special operations forces, intelligence, and military police. . . . Beginning in fiscal year 2006, the Army will undertake a major rebalancing effort involving over 80,000 spaces to further relieve stress on the force and continue to improve its Reserve component capabilities and readiness. (Rumsfeld, 2003, pp. 13–14)

Restructuring the Army

The rebalancing of spaces is complemented by an even more radical plan. Employing the new force generation concept, the Army plans to restructure itself to become an "expeditionary" force so that it can provide a continuous supply of forces more effectively than it has in the past. The Secretary of the Army and the Chief of Staff of the Army recently wrote about their plans for changing the Army—plans that when implemented will finally address the changed state of the world since the fall of the Soviet Union and radically transform the Army from the Cold War force that has lingered for the past 15 years to a force designed to address the realities of the post–Cold War environment. In the summer 2004 issue of the Army's senior professional journal, *Parameters*, the Acting Secretary of the Army and the Chief of Staff wrote:

> In the Cold War, the United States was committed to reinforce Europe with ten divisions within ten days, but no one perceived that responsiveness as expeditionary. The reason for this is significant: in the Cold War we knew where we would fight and we met this requirement through prepositioning of units or units set in a very developed theater. The uncertainty as to where we must deploy, the probability of a very austere operational envi-

> ronment, and the requirement to fight on arrival throughout the battlespace pose an entirely different challenge—and the fundamental distinction of expeditionary operations. (Brownlee and Schoomaker, 2004, p. 9)

To meet the new environment it now faces, the Army is adopting the deployment cycle strategy that the Navy and Marine Corps have used for years, as well as the Air Expeditionary Force concept more recently adopted by the Air Force. Instead of having all active combat units of the Army constantly at a high state of readiness and all available to deploy, only one-third of the active combat units of the Army will be available to the president at any time. Moreover, only one-sixth of reserve component troop units will be available to be deployed at any point in time.

Michael O'Hanlon, a senior fellow at the Brookings Institution, however, has noted what may be the critical weakness in the expeditionary model the Army is moving toward. In *Parameters*, O'Hanlon posed the central question, "How does one determine the appropriate . . . size of the Army?" and brought the discussion back to the very heart of the Army's plan. He answered his own question:

> There is no definitive method because it is impossible to determine exactly how large a rotation base will be needed to continue the Iraq mission over a period of years while avoiding an unacceptable strain on the all-volunteer force that could drive large numbers of people out of the military. (O'Hanlon, 2004, p. 10)

The Simple Math of Rotation and Deployments. In July 2004, the Chief of Staff of the Army reported that there were 33 active and 15 National Guard brigades that were "relatively immediately available" (Schoomaker, 2004). To reduce the frequency with which units are called and to provide a higher degree of predictability in soldiers' lives, GEN Peter Schoomaker said,

> [W]e have a force generation model that shows how we can turn the active force on about [a] 3-year rotation, always having adequate brigades available to us, and turn the Army National Guard

> and the Army Reserve on a five-to-six year rotation that would likewise provide us with a predictable flow of available units. (Schoomaker, 2004)

The "simple math" of unit rotation, however, suggests that the path the Army has set upon may not meet its stated needs. The Army's current situation is not robust, and the 1:3 ratio for the active force and 1:5–6 ratio for the reserve force is not consistent with maintaining the current deployment level of 16 brigades. To maintain 16 brigades in theater for one year out of three years requires a force of 48 brigades, not the planned 43 brigades.[21] Although the active Army is currently moving from 33 brigades to 43, once achieved, it will still be five brigades short. The National Guard can take up some of the slack, but problems still remain.

While GEN Schoomaker acknowledged that there were 36 National Guard brigades, "only 15 of them were resourced at an increased level" (Schoomaker, 2004). Given a once-every-six-year rotation and deployment schedule, and calling a National Guard brigade for about 18 months—a three-month redeployment, a 12-month deployment, and a three-month stand-down—it would take two National Guard brigades to provide the same coverage as one active Army brigade.[22] The 15 "resourced" National Guard brigades could provide the same level of coverage as about seven active brigades. In other words, the Army starting with about 40 active/National Guard brigades would build to about 50 active/National Guard brigades, against a requirement for 48 brigades.

---

[21] The simple math is 16 x 3 = 48, not 43.

[22] An 18-month mobilization is not in complete accord with Secretary Rumsfeld's stated policy of mobilizing a reserve unit no more than one year out of six years. A 12-month mobilization, however, is not practical if a National Guard brigade needs to take a turn in the current 12-month rotation cycle. If pre- and postdeployment time were taken out of the 12-month call-up, only six months would be available to be deployed. In terms of covering deployments, it would take four National Guard brigades rather than two to equal each active Army brigade. The extra six months every six years would increase the "deployment effectiveness" of the National Guard by 100 percent; decrease the training and deployment costs by 50 percent; and add 50 percent to the PERSTEMPO burden but only 8 percent to the time-away-from-home of the 72-month cycle.

Transforming the Air Force, Navy, and Marine Corps

It has not been only the Army that has had to change the way it does business to meet the stress of the new post–Cold War environment. Even before September 11, 2001, the Air Force and Navy were heavily engaged in the Balkans, Iraq, and the Persian Gulf.

Air Force. During the 1990s, the phrase "low density, high demand" was heard many times in the Pentagon to describe situations in which military police units, psychological operations units, or electronic jamming aircraft and their crews, to name just a few, were constantly on deployment. This was because there was a great demand for their services, but fewer units than were needed had been provided in the active forces. Such jobs, it was argued, should be left to the reserve forces, but even there an inadequate number of units had been provided. Lt Gen Lawrence P. Farrell, Jr., Air Force Deputy Chief of Staff for Plans and Programs, had the following viewpoint:

> The problem is that since about 1990, we found ourselves continuing to rotate forces to enforce the protocols from the desert war and for other purposes. We got involved in Northern Watch and Bosnia and, without really realizing it, we found ourselves in a series of ongoing, expeditionary operations. . . . We have been approaching such deployments on what amounted to an ad hoc schedule basis . . . . Recent USAF quality-of-life surveys confirm that the impact of deployments has been almost as severe on some of the support specialists at domestic bases as on the overseas participants. Moreover, the polls show a close connection between increased optempo and falling retention rates. (as quoted in Callander, 1998)

The Air Force's answer to these problems was to "structure the forces into standing units, and in peacetime they would train together, plan together. . . . Then, when their turn came to go on deployment, they would know a year ahead so they could plan on it" (Callander, 1998). In 2003, Air Force Chief of Staff Gen John P. Jumper, an early proponent of the concept when he was Central Command Air Forces commander in the late 1990s, told the Air Force Association about the benefits of the Aerospace Expeditionary Force (AEF) in managing the all-volunteer force: "The AEF is allowing us to highlight our stressed career fields. We are able to pinpoint them and able to size the level of our stress. . . . We are working hard to right-size our force" (Jumper, 2003).[23] In 2004, Gen Jumper reported to the House Armed Services Committee

> that while reconstitution of air expeditionary forces is not moving as quickly as expected, the concept is battle proven. . . . 'We have extended our deployment time from 90 to 120 days, [and] we have about 8 percent of our force on 120-day rotation. About 20 percent of the high demand forces are on rotations [lasting] up to one year.'" (as quoted in Lopez, 2004)

Making the current situation in Iraq somewhat easier for the Air Force than it is for the Army, the Air Force is trying to reduce its end strength as it continues to shrink in size. As a result, Gen Jumper could report, "We are enjoying excellent results in our recruiting and retention" (as quoted in Lopez, 2004).[24] In addition, the Air Force has had success in rebalancing the force by moving airmen from career fields with overages into career fields with shortages.

Navy. As the Army and the Air Force incorporate the theme of deployments and rotation into the way they do business, they are moving toward the model the Navy has used for decades. With the vast majority of the fleet homeported in the United States, the Navy has cycled its ships and aircraft to foreign waters on well-established and predictable cycles, usually a six-month deployment followed by 24 months at home (Pike, 2005). In times of emergency, the fleet could surge. During Operation Iraqi Freedom, 50 percent of Naval forces were forward deployed, including seven carrier strike groups (CSGs) and eight large-deck amphibious ships. In FY 2003, the Navy introduced the Fleet Response Plan to provide "the nation with increased naval capabilities and more deployment options" (England, 2004). When fully implemented, the period of time between

---

[23] For a history of the Air Expeditionary Force concept, see Titus and Howey (1999).

[24] For example, in September 2005, the Air Force achieved 103 percent of its non-prior-service accession goals and "exceeded their annual retention goals" (Assistant Secretary of Defense [Public Affairs], 2005).

deployments—the Inter-Deployment Readiness Cycle—will increase from 24 months to 27 months, but the readiness of ships in the period between deployments will also increase. Maintaining a higher level of readiness while at home means that the Navy will have six to seven CSGs *employable*—deployed or capable of being deployed—rather than its traditional three or four.

As the Navy moves to increased readiness, some have expressed concern that the higher levels of readiness will increase the workload and stress the crews experience between deployments, even as the length of time between deployments decreases. The usual concerns, however, about possible lower retention or fewer willing to join the Navy seem less a problem because the Navy is moving to fewer ships and aircraft and is reducing the number of sailors in the force. As in the early 1990s, the Navy is reducing accessions and employing a number of other programs to "shape" its force, including, among other things, an Involuntary Release from Active Duty program, a Selected Early Retirement program, and High Year Tenure, employing a more stringent up-or-out system by reducing force-out points from 10 years of service (YOS) to 8 YOS for E-4s and from 20 YOS to 14 YOS for E-5s.

Marine Corps. Current operations in Iraq are stressing the Marine Corps. According to the Defense Manpower Data Center (DMDC), with roughly 25,000 troops in theater, 351 marines had been lost to enemy action by January 29, 2005 (DMDC, 2005), a casualty rate roughly 2.5 times greater than the Army's. While the smaller Marine Corps, like the Navy, was a force built around deployments to Okinawa, and rotations to the Fleet Marine Forces for six-month deployments on amphibious ships, sending so many troops to Iraq—approximately 35,000 deployed in combat operations worldwide and 25,000 in Iraq for a seven-month deployment (Lisbon, 2004)—the Marine Corps developed new ways to manage its force. Specifically, the Corps has developed "provisional units" to ensure that the critical skills needed in Iraq are available.

For example, of the nine Marine Corps infantry battalions in Iraq in the summer of 2004, one was from the reserves, and "other support units, as well as individual reservists augmented the active-duty force"

(Lisbon, 2004). In total, approximately 20 percent of the Marine forces in theater were reservists. To man such high-demand, low-density military occupations, the Marines have used Individual Mobilization Augmentees and increased Selected Reenlistment Bonuses, permitted unlimited cross-year extensions, and accepted volunteers from the Individual Ready Reserve (IRR) or retirees. The Marine Corps ruled out using "stop losses" to stabilize units deployed to Iraq or involuntarily activating the IRR (Lisbon, 2004).

In the short run, to increase manning with the operating forces, Congress allowed the Marine Corps to increase its end strength by 3,000, to 178,000. The Marine Corps is restructuring its force to better meet the demands of the U.S. global war on terrorism. Upwards of 4,000 military jobs will be converted to civilian positions. Additional infantry battalions will be created with increases in high-demand, low-density areas—ordnance disposal, intelligence, and so forth. The active/reserve mix is also being changed. In the future, active-duty units will include those with skills that had traditionally been used only in the reserves, such as civil affairs (Rhodes, 2004).

Family Program to Ameliorate the Most Negative Aspects of Deployment

There is more to managing the force than just compensating people for their service or organizing the force to make sure that it can best meet current demands. Providing support services for service members and their families helps ameliorate the most negative aspects of deployments. One of the consequences of the all-volunteer force is that the size of the career force substantially grew, and the force was much more likely to have families (i.e., dependents) than did the mixed force of volunteers and draftees of the 1960s. When the Gates Commission started to examine the possibilities of moving to an all-volunteer force, 27.4 percent of the active force had served for more than four years; by June 1977, that number had grown to 41.4 percent of the force (Wisener, 1979, Table K-14). In 1971, 44.4 percent of the active-duty enlisted force had dependents. By the end of June 1977, that figure had grown to 51 percent (Wisener, 1979, Table K-12). The largest increases were in the Army. Over the same period, Army enlisted members serving more than four years grew from 20.3 percent in 1970 to 36.8 percent in 1977, and enlisted members with dependents increased by 30 percent, from 38 percent of the enlisted force to 49.4 percent of the enlisted force. In September 2004, 56.5 percent of Army enlisted personnel had families (DMDC, 2004). Even in the Marine Corps, with its emphasis on youth and first-term personnel, marines with families made up 41.0 percent of the enlisted force, compared with 19.5 percent in 1970. The number was higher for officers, with 69.4 percent of the officer corps married.[25] But it was not always that way.

The Traditional Army

Traditionally, military life has not been "family friendly," as made clear by one account of life in the British Army in the 1850s:

> In most barracks, the men eat and drink in the same rooms which serve them as dormitories . . . . Some of the soldiers are permitted to have their wives with them, but no suitable arrangements are made for their indulgence. It has recently been ascertained that, in 251 barracks, no less than 231 were without separate accommodation for married soldiers; the women (a few in each company) resided with husbands under circumstances repulsive to every sense of delicacy and propriety, and even in the exceptional instances, the space afforded to an entire family is not more than ought to be allowed for a single individual. (*Chambers's Journal* editorial staff, 1859a)

In America, things were no better. The American Army first took note of service members' families in 1794, when a death allocation of cash was designated to "widows and orphans of officers killed in battle" (Department of the Army, undated). The "benefit" was later extended to the families of NCOs. As in Great Britain, married soldiers were expected to provide for their family's needs. "Wives, known as 'camp followers,' could receive half-rations when they accompanied their spouse and performed services such as cooking, sewing, cleaning

---

[25] Data from DMDC (2004).

barracks, working in hospitals, and even loading and firing muskets" (Department of the Army, undated). In 1802, the Army authorized company laundresses, many of whom married NCOs. By regulation, however, the Army barred officers from marrying until their captaincy; NCOs and enlisted men required permission from their company commander to marry. In 1847, Congress prohibited the enlistment of married men in the Army.

After the Civil War, the Army followed a policy of discouraging married men from serving. In an effort to reduce the number of families, the Army provided family quarters only for senior officers. Married men could not enlist, and the Army provided little assistance to service members with wives and children. The Army did not provide housing for married enlisted men's families, did not provide transportation for their family when a soldier permanently changed his duty station, and "obstructed" the reenlistment of married soldiers. Whatever support the families of married enlisted personnel got came from the largesse of the wives of officers and NCOs (Department of the Army, undated). Until World War II, with the exception of the period of World War I, the adage, "If the Army had wanted you to have a wife, they would have issued you one," aptly summed up the service's attitude toward families.

During World War I, while most married men were not drafted, the government still had to provide support for those who were. As a result,

> World War I . . . ushered in the first program of family allotments for officers and enlisted personnel, voluntary insurance against death and disability, and other family assistance measures. On the eve of World War II, Congress furnished government housing for soldiers E-4 and above with family members. After the start of hostilities the Army issued a basic allowance for quarters for military families residing in civilian communities. With the exclusion of married men from the service no longer feasible, the Army granted monthly family allowances for a wife and each child. Married females, on the other hand, were barred from enlistment and could be separated from the service because of pregnancy, marriage, and parenthood, a policy that remained in effect until 1975. To deal more effectively with family emergencies, the

> Secretary of War created the Army Emergency Relief (AER) in February 1942. The AER adopted the slogan: "The Army Takes Care of its Own." (Department of the Army, undated)

The Cold War–era Army in no way resembled the pre–World War II organization of the same name. The postwar Army was many times the size of the prewar Army and had worldwide responsibilities. "The Army's approach to addressing family concerns[, however,] remained reactive and piecemeal. The development of the Army Community Services (ACS) organization in 1965 [at the start of the buildup for Vietnam] was the Army's first attempt to create an umbrella approach for family support" (Department of the Army, undated). But it took the move to the all-volunteer force to really change things.

Recruiting Soldiers and Retaining Families: The Development of Army Family Programs in the All-Volunteer Force

On the eve of the all-volunteer force, the *Fiscal Year 1971 Department of the Army Historical Summary* made no mention of military families per se; it was only implied by the concern that "the Army needs a total of 353,440 housing units for eligible families [when] available family housing on and off post total 220,600 units" (Bell, 1973, p. 55). By 1978, however, the Army understood that its approach to its Quality of Life program, originally established to "improve services and activities for enlisted personnel in their daily life," needed to be expanded "to bolster [a] community of life support activities" (Boldan, 1982, p. 91). Citing the all-volunteer force, the Army noted before the end of the draft that

> less than half of the soldiers were married. By the end of 1977, over 60 percent fell into that category, many more were sole parents, and a considerable number were married to other soldiers. The changing composition of the Army necessitated increased attention to community services to sustain morale and retain highly qualified personnel. (Boldan, 1982, p. 91)

In 1979, the Army, in recognition that even the most junior enlisted members had families, established "a family separation allowance for service members in grades E-1 to E-4" (Brown, 1983, p. 10).

In March 1980, the Deputy Chief of Staff for Personnel told Army families the following:

> Our commitment to the Army family has been made at the highest level. We know that the Quality of Life impacts on readiness and on attracting and retaining quality soldiers the Army needs. We've got to continue to get better in this vital area, and through our efforts, provide meaning to the [resurrected World War II] slogan: "The Army Takes Care of Its Own." (Brown, 1983, p. 92)

Specifically, the Army had begun making a major commitment to child care programs. By FY 1980, the Army had 281 child care programs (159 day care and 122 preschool) in operation (Brown, 1983, p. 92).

In October 1980, the first Army Family Symposium was held, in Washington, D.C. Sponsored by the Army Officers' Wives Club of the Greater Washington Area and the Association of the United States Army, nearly 200 delegates and observers attended. The symposium resulted in the creation of the Family Action Committee. Following the symposium's recommendation, the Chief of Staff of the Army established the Family Liaison Office within the Office of the Deputy Chief of Staff for Personnel to oversee all family issues. On September 8, 1981, the Adjutant General's Office opened the Army Family Life Communications Line at the Pentagon and developed a quarterly family newsletter to be distributed to Army families worldwide (Hardyman, 1988, p. 109).[26]

On August 15, 1983, Army Chief of Staff John A. Wickham signed the *Army Family White Paper—The Army Family*, which has been described as a "landmark document [that] underscored the Army's rec-

---

[26] Attitudes were also changing, as the *Army Historical Summary* notes: "The Chief of Staff also directed the general use in Army publications of the terms family member or spouse in place of dependent, and he issued a policy statement supporting the right of family members to be employed without limiting a service member's assignment or position in the government. The policy statement read in part: 'The inability of a spouse personally to volunteer services or perform a role to complement the service-member's discharge of military duties normally is a private matter and should not be a factor in the individual's selection for a military position'" (Hardyman, 1988).

ognition that families affect the Army's ability to accomplish its mission" (Department of the Army, undated). It provided for the annual *Army Family Action Plan*, the Army theme for 1984 ("Year of the Family"), and the enhancements of installation-based Family Centers and the creation of the U.S. Army Community and Family Support Center, which combined the Army's nonappropriated fund activities and the family office into a single command headed by the former Adjutant General of the Army, Major General Bob Joyce. In addition, the Army's Vice Chief of Staff, General Max Thurman, pushed for additional child care centers, which were important in supporting the growing numbers of military wives and female service members, a further result of the move to an all-volunteer force.

With the end of the Cold War and America at peace, issues the delegates brought to the Army Family Action Plan (AFAP) conference in the early 1990s[27] seemed rather mundane compared with those of a decade earlier: "inadequate housing allowances, comprehensive dental care, and enhanced family programs for the Total Force were among top issues identified in 1990. Inequitable military pay, the need for increased marketing of CHAMPUS (Civilian Health and Medical Program of the Uniformed Services), and underutilized teen programs were issues identified in 1991" (Janes, 1997). The situation changed, however, when Iraq invaded Kuwait in the summer of 1990.

Operations Desert Shield and Desert Storm: The Volunteer Army Goes to War. Simultaneously, with the initial deployment of troops to Saudi Arabia, and a harbinger of things to come, the ACS established and operated 24-hour Family Assistance Centers (FACs) at the seven stateside posts from which large numbers of troops deployed. The FACs brought together chaplains, lawyers, relief workers, and other social service specialists "under one roof" to provide information and counseling. The ACS trained "unit support groups" and provided relocation information, consumer and financial advice, employment counseling,

---

[27] It should be noted that "[i]n November 1989 Congress passed the Military Child Care Act (MCCA). This legislation stipulated minimum appropriated funding and staff levels, higher wages, and better training for child care staffs; user fees based on family income; national accreditation of child development centers; and unannounced inspections of local child development services (CDS) programs and facilities" (Janes, 1997).

aid to exceptional family members, and other services. Unit support groups, and traditional support groups such as the United Services Organization, the American Legion, the YMCA, and the American Red Cross, provided information and helped with child care, housing, and financial issues. From August 1990 through January 1991, the Army Emergency Relief helped 31,000 soldiers and their families with $17 million in grants and interest-free loans (Janes, 1997).

Soon after the first troops started to flow to the Middle East, the Army, in August 1990, established a toll-free hotline staffed at an operations center in Alexandria, Virginia, to support the reserve components and those families at installations without FACs. The hotline was staffed 24 hours a day/seven days a week through April 1991 and then went to reduced hours until July 1991. The center logged 80,000 calls during the nine-month period.

For the Army, the lessons of Desert Storm were that "[f]amily members of deployed service members had innumerable problems and questions, felt confused and abandoned, and often did not know where to turn to obtain resolution and answers" (Reeves, 1998). To address these needs and to "create self-sufficient and self-reliant individuals and families who could cope with the stress of deployment," the Army developed the Family Team Building Program. Given that, by the mid-1990s, 66 percent of Army personnel were married, 54 percent of spouses were working, and an additional 8 percent were single parents, it should have come as little surprise when the Army established family support groups as a major source of support for every deployment and declared that "[q]uality of life is the Army's third highest priority, immediately behind readiness and modernization" (Reeves, 1998).

Deployments in the 1990s

After the Gulf War and throughout the 1990s, ever-increasing deployments placed new demands on soldiers and their families that largely did not exist during the Cold War. Army Chief of Staff Dennis Reimer told Congress, "Deployments to places such as Bosnia, Kuwait, Haiti, Honduras, the Sinai, Macedonia, and elsewhere mean that on any given day, the Army commits the resources of approximately four divisions" (West and Reimer, 1997). While the manpower in the Army was down 3 percent since the end of the Cold War, Reimer reminded Congress that deployed operations were up some 300 percent. He estimated that "officers and senior non-commissioned officers from deployable units now spend 180–190 days away from home annually, while junior soldiers spend 140–155 days away" (as recounted in Ryan, 1998, p. 3). Reimer also commented, "As units deploy more frequently Army families must be prepared to deal with the stress and uncertainty that deployment brings. The Army strongly supports family programs that prepare soldiers, Army civilians, and their family members for separations" (West and Reimer, 1997). During Operation Joint Endeavor in Bosnia, the U.S. Army, Europe, activated 21 FACs to serve as one-stop centers for families to obtain information and services. These centers also afforded deployed soldiers the peace of mind of knowing that their families were being cared for.

The Army's support for family programs and concerns for the impact that frequent deployments might have on the all-volunteer force was a reflection of the survey responses the service received from troops. During the height of the deployments to Bosnia in the late 1990s, survey results showed that

> [s]oldiers intending to leave the military also were more likely to report that the number of deployments had hurt their marriage and caused a strain on their family than those soldiers who reported that they were remaining in the military. However, even for those soldiers who reported that they would stay until retirement, about half reported that deployments had put a big strain on their family.[28] Thus, although deployment tempo appears to take its toll on soldier retention and family well-being, it has the potential to work as a motivational force as well. (Castro and Adler, 1999)

---

[28] "For example, among the 1,305 soldiers with families surveyed, 61.7 percent of those intending to get out after their obligation reported that the number of deployments caused a big strain on the family versus 54 percent of those intending to stay past their obligation, and 49 percent of those intending to stay at least until retirement" (Castro and Adler, 1999).

Despite increased deployments, in February 1997 the Army reported to Congress that it had "accomplished 100 percent of its initial-term and mid-career reenlistment goals" (West and Reimer, 1997).

Operation Enduring Freedom and Operation Iraqi Freedom

On February 5, 2003, the Army's 2004 Posture Statement started with these words: "Our Nation, and our Army, are at war" (Brownlee and Schoomaker, 2004). Compared with the posture statement of just a year earlier, this was a remarkable change. In stark and sober tones, the 2004 Posture Statement uses the word "stress" five times to describe the current situation when the word was not used at all the previous year: "Operation Iraqi Freedom and Operation Enduring Freedom have stressed the force" (Brownlee and Schoomaker, 2004, letter); "As a result of this adaptive enemy and worldwide commitments . . . [we] will continue to be stressed" (p. 2); "These deployments, coupled with planned future rotation(s) . . . have highlighted already existing stress to our force" (p. 3); "We will continue ongoing efforts to restructure our forces in order to mitigate stress" (p. 14); and "Our army will remain stressed to meet anticipated requirements" (p. 14).

Contemporary Programs to Mitigate Stress

In April 2002, DoD published *The New Social Compact*. The heart of the document was a reciprocal understanding between the department and service members and their families. The document declared: "Service members and families together must dedicate themselves to the military lifestyle, while the American people, the President, and the Department of Defense must provide a supportive quality of life for those who serve" (Molino, 2002, p. 1). The compact provided an "overview of services' delivery systems and strategies" (Molino, 2002, pp. 103–113). The overview noted that "[t]he apparent similarities and differences between these program approaches demonstrate that supporting families can be accomplished in several ways" (Molino, 2002, p. 103).[29]

---

[29] The Army Well-Being program noted above was one of the programs discussed, along with the Navy and Marine Quality of Life Master Plan and Navy Lifelines; the Air Force Quality

The National Defense Authorization Act for Fiscal Year 2003 required that "[t]he Secretary of Defense shall every four years conduct a comprehensive examination of the quality of life of the members of the armed forces (to be known as the 'quadrennial quality of life review')." The *1st Quadrennial Quality of Life Review (QQLR), Families Also Serve,* was issued in May 2004. The document commits DoD to working "hard to help military families deal with the *stress* attributable to separations and a range of *uncertain* war-time conditions" (DoD, 2004, emphasis added). In response to a request from Congress,[30] the report highlights "Family Centers & Unit-Based Support of Mobilization," "Deployment," and "Return/Reunion." The report highlights the work of various installation Family Centers and local FAC groups that support active-duty service members and their families, as well as the "Joint Family Readiness Working Group . . . [and the] approximately 400 National Guard Family Assistance Centers" (DoD, 2004, p. 85).

Military OneSource Program and Family Assistance Counseling. Of particular importance to deploying service members and their families is on-call counseling for families in distress. Noting that "[t]he Navy is currently the only Service with professional counseling services in family centers" (DoD, 2004, p. 63), DoD found "[t]he lack of counseling services to assist troops and families in coping with stress results in increased family deterioration, frequent duty disruptions, and dissatisfaction with military life. This situation, in turn, negatively impacts unit readiness and compounds retention problems" (DoD, 2004, p. 63).

---

of Life, Community Capacity Model, and Air Force Crossroads; and the Reserve/Guard Component Family Support Initiatives (Molino, 2002).

[30] The Senate Appropriations Committee Report 108-33 notes, "Personnel Support—The Committee strongly supports the members of our Armed Services and their families. In particular, the Committee notes that those now deployed in current military operations, as well as the members of our National Guard and Reserve who have been called to active duty, have been activated frequently and for extended periods of time since September 11th. The Committee urges the Secretary of Defense to identify those hardships experienced by service members and their families and propose remedies to address those hardships. The Committee directs the Secretary of Defense to provide a report to the Congressional Oversight Committees, no later than July 1, 2003, on how the Department can better support our deployed service members and their families" (Senate Appropriations Committee, 2003).

To help address this problem the QQLR publized the Department's implementation of Military OneSource, designed to deliver information and referral services to troops and families worldwide and to service the many families that do have ready access to Family Service Centers. The QQLR summarized the role of the Military OneSource program this way:

> Military OneSource is available 24/7. 365 days of the year . . . by calling 1-800-342-9647 families can obtain information on, among other subjects, child care, parenting, housing and education, budgeting and medical services, at any location, world wide. Military OneSource takes support services to all members of the Armed Forces, including the Reserves and National Guard members and families who do not live on military installations and often can't take advantage of what DoD has to offer them. Military OneSource is an augmentation to, not a replacement of, the installation family centers. Each of the Military Services will have fully implemented this service by the end of FY 2004. (DoD, 2004, pp. 62–63)

Other Support Programs. Each of the services has developed programs to mitigate stress. The Army's program is called "Well-Being." It uses the Internet, which has become one of, if not the most important, way the current generation of soldiers and their families communicates and learns about things that can help their lives.

At the top of the Army home page (http://www.army.mil/) is a link to the Army's Well-Being program. (Figure 3.1 shows the Army's Well-Being home page.) The Army Well-Being program provides the umbrella under which a full range of support programs is housed. On the Well-Being Web site, the link to family programs is the gateway to a myriad of support programs for Army soldiers and their families. Figure 3.2 shows the Army Well-Being link to Family Programs and a number of the programs that are available that are not specific to the bases throughout the world.

Figure 3.1
Army Well-Being Home Page

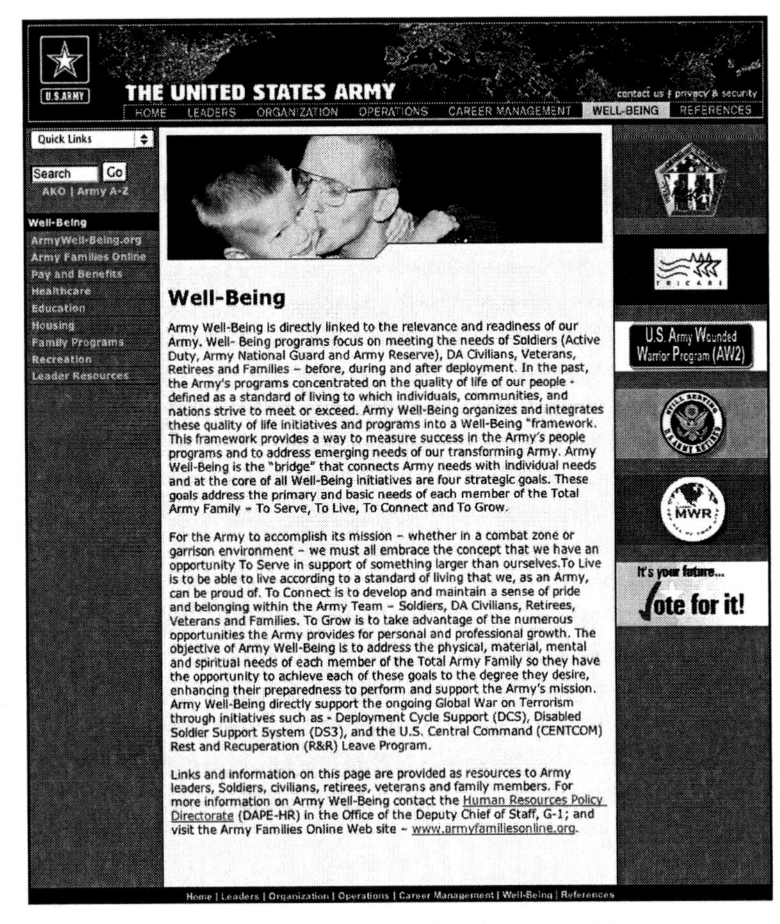

SOURCE: http://www.army.mil/wellbeing/ (as of August 2006).

The Navy offers the Lifelines Service Network over the Internet. Each of the services offers support programs to families on base. Besides the Army's ACS centers, there are the Air Force Family Support Centers, the Marine Corps Community Service Centers, the Navy Fleet and Family Support Centers, and the Coast Guard Work-Life Centers. The National Guard and reserve components also have programs. The ombudsman program, unique to the Navy and Coast Guard, is an information link between unit commanding officers and the families

Figure 3.2
Army Family Programs Linked to the Army Internet Home Page

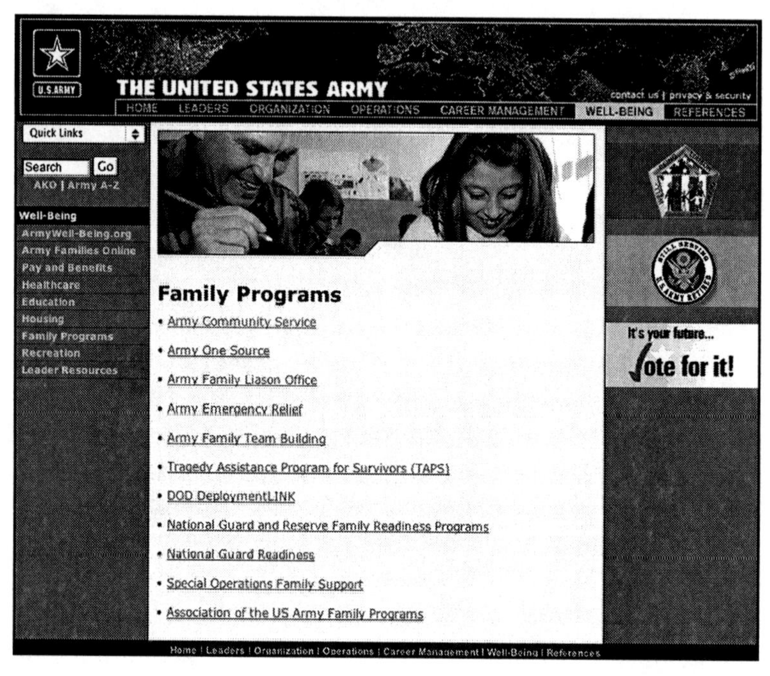

SOURCE: http://www.army.mil/wellbeing/family.html (as of August 2006).

of their personnel. A similar program, called key volunteers, operates in the Marine Corps. The Armed Services YMCA is a social service agency with branches and affiliates at many U.S. installations. Armed Forces Emergency Services of the American Red Cross provides a vital link home to those serving in remote areas. Private military relief societies can help service members solve emergency financial problems.

There are also four private, nonprofit financial aid societies: Army Emergency Relief, the Air Force Aid Society, the Navy–Marine Corps Relief Society, and Coast Guard Mutual Assistance. Each has local representatives on military installations, often in Family Centers. Even with so many programs, it is hard to know which ones work and which ones do not work, and under what circumstances.

Effectiveness of Military Family Support Programs

From the very beginning of the modern family program, policymakers have been asking for some level of proof that family support programs are cost-effective. The *Department of the Army Historical Summary* for FY 1981 noted that "[t]he Quality of Life Program, after three years of planning and programming, at last received enough funds to make a noticeable difference for soldiers and their families" (Hardyman, 1988, p. 108). With costs projected to run $1.6 million over the next six years, the *Summary* noted: "Quality of life efforts have been handicapped in the competition for limited resources by the Army's inability to quantify the benefits derived from implementing the initiatives. There was no obvious way to measure soldiers' satisfaction and its effect on soldier commitment" (Hardyman, 1988, pp. 108–109). With a sense of hope, the *Summary* told of the Army's hiring of "a consulting firm to develop a model to forecast the effects of quality of life initiatives and the necessary levels of funding to achieve the greatest improvement in retention" (Hardyman, 1988, p. 109). The Army hoped it would have its forecasting model by April 1983. In 2004, the *1st Quadrennial Quality of Life Review* reported that, despite the general recognition that quality of life "impacts the retention of service members and the readiness of the armed forces, . . . research that can inform policy on these issues is surprisingly inadequate" (DoD, 2004, p. 187).

There are several meanings to the phrase "inform policy" as used in the 2004 *Review*. At one level, some policymakers continue to ask, as they did in 1981, what they are getting for the money spent on these programs. As noted in the *Review*, there is a general acceptance, mostly based on anecdotes, that quality of life affects the retention of service members and the readiness of the armed forces. At another level, it is the individual programs that need to be assessed to determine what is and is not working. A recent study by the National Military Family Association confirmed that "[m]any programs and services are in place to help military families, [but concluded that] these programs, however, are inconsistent in meeting families' needs" (Wheeler, 2004, p. 8).

What Do We Know About the Effectiveness of Family Programs?
Today, surveys and focus groups are the primary means we use to

learn about these programs, but they provide an incomplete picture. Academic research that focuses on how people make the decision to stay or leave also provides little insight into where DoD should spend its money.

Surveys. Surveys provide a great deal of what we know about overall attitudes toward family support programs. In November 2003, the DMDC, using the Status of Forces Surveys of Active Duty Personnel, questioned 11,546 service members from all the services and of all ranks. One-third of the sample was from the Army; 54 percent of the sample were enlisted personnel. The vast majority (93 percent) had some family responsibilities, either being single with children (24 percent) or married. Half the sample was "married, with children," and only 7 percent was "single, without children" (Survey & Program Evaluation Division, 2004). The survey contained the following questions about support services:

- Did you receive support services (e.g., support groups, counseling, pre- or postdeployment information briefings) before or after returning to your permanent duty station?—75 percent of Army respondents said yes.
- Did the support services (e.g., support groups, counseling, pre- or postdeployment information briefings) help you adjust to returning to your permanent duty station?—58 percent of Army respondents said yes.
- Did the support services (e.g., support groups, counseling, pre- or postdeployment information briefings) help you adjust to returning to your spouse or significant other?—55 percent of Army respondents said yes.
- If you begin to experience difficulty adjusting to returning to your permanent duty station and/or family life, do you know where to go for help?—86 percent of Army respondents said yes.
- Would private personal or family counseling be useful to you or your family?—39 percent of Army respondents said yes.

While the questions are helpful in getting a general understanding of how service members view support services, they do not provide enough detail about the needs of members or their families, or their experiences with support services to make critical decisions about how best to manage the program. The survey does not help managers understand why 25 percent of the Army apparently did not receive support services or, of the approximately one-third of those who did, why they did not feel the services helped them either adjust to work or help them with their spouse or significant other.

Focus Groups. Focus groups can complement surveys and may develop explanations for some of the issues that surveys are unable to address. For example, a recent RAND study attempted to explain the seemingly anomalistic behavior of deployed soldiers who were more likely to reenlist than similar soldiers who were not deployed (Hosek, Kavanagh, and Miller, 2006). The researchers thought that focus groups with military personnel who had and had not been deployed would allow them to "ferret" out possible explanations, providing new perspectives on the reenlistment decision. Focus groups, however, largely stay in the realm of anecdotal information. They are not intended to provide a representative sample, and therefore the "insights" they provide cannot be rigorously tested. However, focus group are very useful as one way to develop "hypothesis" and guide further inquiries.

Questions about program effectiveness have endured since the early days of the all-volunteer force, but progress toward answering these questions has been very slow, which suggests how difficult this problem has turned out to be. Problems persist in determining the correct sampling design and the analytic and statistical approaches to follow. Overdue is a valid and reliable research design for the collection and analysis of information to assess the performance of the variety of family support programs.

CHAPTER FOUR

Summary and Conclusion

Headlines notwithstanding, the all-volunteer force has done extremely well during these stressful and uncertain times. Commissioner Greenewalt's certainty in 1970 that "[a]n armed force involved in a major conflict could *not* be voluntary" (Greenewalt, 1969, emphasis in the original) has been proven wrong. History suggests that the conditions favorable to conscription—overwhelming support for the cause and equality of sacrifice—are not present today. The senior leaders in the administration and many in Congress are of an age at which former Secretary of Defense Casper Weinberger's words in 1987—"We know what the draft did to the social fabric of this country in the '60s" (as quoted in Chambers, 1987, p. 259)—are fair warning. The American military has been very resilient in finding ways to make the all-volunteer force work. However, a number of new and expanded compensation programs have been put in place and retention has remained high; each of the services has restructured to provide additional personnel to meet the demands of new missions; and family programs have been expanded to mitigate stress.

As it has been from the beginning, the all-volunteer force remains fragile. Accordingly, DoD has provided a wide range of support programs to help service members and their families cope with the stress and uncertainty of heightened military operations and deployments. To date, increases in the operational tempo for active and reserve forces, including multiple tours in the combat areas of Afghanistan and Iraq, have not resulted in significant recruitment shortages or low retention. However, only time will tell.

References

105th Congress, *A Statement of the Posture of the United States Army: Fiscal Year 1998*, hearing before the First Session, February 1997.

108th Congress, *A Statement of the Posture of the United States Army: Fiscal Year 2004*, hearing before the Second Session, February 5, 2004.

Allard, Kenneth, *Somalia Operations: Lessons Learned*, Washington, D.C.: National Defense University Press, 1995.

Assistant Secretary of Defense (Force Management and Personnel), *Special Separation Policies for Survivorship*, Washington, D.C.: Department of Defense, 2003.

Assistant Secretary of Defense (Manpower and Reserve Affairs) (ASD[M&RA]), *Enlistment Attraction and Retention Incentive Pays: Report of the First Quadrennial Review of Military Compensation*, Washington, D.C.: Department of Defense, 1967a.

———, *Report of the First Quadrennial Review of Military Compensation*, Washington, D.C.: Department of Defense, 1967b.

———, *Special Pays—Hostile Fire Pay: Report of the First Quadrennial Review of Military Compensation*, Washington, D.C.: Department of Defense, 1967c.

Assistant Secretary of Defense (Public Affairs), *DoD Announces Recruiting Retention Numbers for September*, Washington, D.C.: Department of Defense, Office of the Assistant Secretary of Defense (Public Affairs), 2005.

Associated Press, "Army Manpower Draft Expected to End in June," *New York Times*, August 5, 1949.

Babington, Charles, and Don Oldenburg, "House GOP Brings up Draft in Order to Knock It Down," *Washington Post*, October 6, 2004.

Bell, William Gardner, ed., *Department of the Army Historical Summary: Fiscal Year 1971*, Washington, D.C.: U.S. Army Center of Military History, 1973.

Boldan, Edith M., ed., *Department of the Army Historical Summary: Fiscal Year 1979*, Washington, D.C.: U.S. Army Center of Military History, 1982.

Bowman, Tom, "Army Reserve Fast Becoming 'Broken' Force," *Baltimore Sun*, January 5, 2005.

Brown, Lenwood Y., ed., *Department of the Army Historical Summary: Fiscal Year 1980*, Washington, D.C.: U.S. Army Center of Military History, 1983.

Brownlee, Les, and Peter J. Schoomaker, "Serving a Nation at War: A Campaign Quality Army with Joint and Expeditionary Capabilities," Parameters, Vol. 34, No. 2, Summer 2004, pp. 5–23.

Bruner, Edward F., *Military Forces: What Is the Appropriate Size for the United States?* Washington, D.C.: Congressional Research Service, 2004.

Burk, James, "The Military Obligation of Citizens Since Vietnam," *Parameters*, Summer 2001, pp. 48–60.

Callander, Bruce D., "The New Expeditionary Force," *Air Force Magazine*, Vol. 81, No. 9, September 1998.

Carr, William, *Briefing: Triple Backstop*, Washington, D.C.: Office of the Under Secretary of Defense (Personnel and Readiness), 2004.

Castro, Carl A., and Amy B. Adler, "OPTEMPO: Effects on Soldier and Unit Readiness," *Parameters*, Autumn 1999, pp. 86–95.

Chambers's Journal editorial staff, "The British Soldier—At Home," *Chambers's Journal*, May 14, 1859b.

———, "The British Soldier—How and Why He Enlists," *Chambers's Journal*, April 29, 1859a.

Chambers, John Whiteclay II, *To Raise an Army: The Draft Comes to Modern America*, New York: The Free Press, 1987.

———, "Conscription," in Eric Foner and John A. Garraty, eds., *The Reader's Companion to American History*, New York: Houghton Mifflin, 1991.

Chief of Naval Operations (CNO), "Force Shaping-Assignment Incentive Pay (AIP) Program," NAVADMIN 161.03, Washington, D.C.: Department of the Navy, 2003.

———, "Revised Selected Reenlistment Bonus and Special Duty Assignment Pay Award Levels for Designated Seal Personnel," NAVADMIN, Washington, D.C.: Department of the Navy, January 27, 2005.

Chief of Naval Personnel Public Affairs, "AIP Program Expands with New Jobs, New Bid Levels," *Navy NewsStand*, 2003a.

———, "Bidding for Billets Could Bring Bundles: Assignment Incentive Pay Proving a Success," *Navy NewsStand*, 2003b.

Chu, David S.C., *Report to Congress: Management of Deployments of Individual Members—Personnel Tempo (PERSTEMPO)*, Washington, D.C.: Office of the Under Secretary of Defense (Personnel and Readiness), 2002.

Clark, Vern, *Joint Tactics, Techniques, and Procedures for Peace Operations*, Washington, D.C.: Joint Chiefs of Staff, 1999.

CNN.com, "Bush Makes Historic Speech Aboard Warship," May 1, 2003, http://www.cnn.com/2003/US/05/01/bush.transcript/ (as of August 15, 2006).

Committee of Conference, *National Defense Authorization Act for Fiscal Year 2004 Conference Report: Section 541—High-Tempo Personnel Management and Allowance*, Washington, D.C.: U.S. House of Representatives, 2003.

Cutler, Frederick Morse, *The History of Military Conscription with Special Reference to the United States*, Worcester, Mass.: Clark University, 1922.

———, "Military Conscription, Especially in the United States," *The Historical Outlook*, Vol. 14, No. 5, May 1923, pp. 170–175.

Dallas Morning News editorial board, "Our Guard Is Down: Iraq War Depends on This Shrinking Force," *Dallas Morning News*, December 21, 2004.

Defense Manpower Data Center (DMDC), *Active Duty Demographics Profile*, Washington, D.C.: Department of Defense, 2004.

———, *War on Terrorism: Operation Iraqi Freedom by Casualty Category with Types as of January 29, 2005*, Washington, D.C., 2005.

Department of the Army, *Consideration of Others (CO2) Handbook*, Washington, D.C.: Office of the Deputy Chief of Staff for Personnel, undated.

Department of Defense (DoD), *Report of the 1st Quadrennial Quality of Life Review: "Families Also Serve,"* Washington, D.C., 2004.

de Tocqueville, Alexis, *Democracy in America,* 1835.

England, Gordon, *Report of the Secretary of the Navy,* Washington, D.C.: Department of Defense, 2004.

Fischer, David Hackett, *Washington's Crossing,* New York: Oxford University Press, 2004.

Flynn, George Q., *Lewis B. Hershey, Mr. Selective Service,* Chapel Hill: University of North Carolina Press, 1985.

———, *The Draft, 1940–1973,* Lawrence: University Press of Kansas, 1993.

———, *Conscription and Democracy: The Draft in France, Great Britain, and the United States,* Westport, Conn.: Greenwood Press, 2002.

Ford, Guy Stanton, "Boyen's Military Law," *American Historical Review,* Vol. 20, No. 3, April 1915, pp. 528–538.

Ford, Worthington C., ed., *Journals of the Continental Congress: 1774–1789,* Washington, D.C., 1904–1937.

"Franco-Prussian War," in *Wikipedia: The Free Encyclopedia,* 2005.

Friedberg, Aaron L., *In The Shadow of the Garrison State: America's Anti-Statism and Its Cold War Strategy,* Princeton, N.J.: Princeton University Press, 2000.

Gallup Brain, "Question qn18 (The Gallup Poll #773) of 1/1/1969," unpublished, 2006a.

———, "Question qn33 (Gallup Poll Social Series) of 10/11/2004," unpublished, 2006b.

Garamone, Jim, "Army's Top NCO Discusses Recruiting, Retention, Optempo," American Forces Press Service, July 26, 2004a.

———, "Defense Act Increases Pay, Provides Benefits," American Forces Press Service, November 1, 2004b.

Gates, Thomas, *The Report of the President's Commission on an All-Volunteer Armed Force,* Washington, D.C.: President's Commission on an All-Volunteer Armed Force, 1970.

General Accounting Office (GAO; now Government Accountability Office), *Peace Operations: Heavy Use of Key Capabilities May Affect Response of Regional Conflicts*, Washington, D.C, GAO/NSIAD-95-51, 1995.

———, *Military Readiness: A Clear Policy Is Needed to Guide Management of Frequently Deployed Units*, Washington, D.C., GAO/NSIAD-96-105, 1996.

Gillert, Douglas J., "Clinton Briefed on Potential Readiness 'Nose Dive,'" American Forces Press Service, September 17, 1998.

Gilmore, Gerry J., "Authorization Act Funds 3.5 Percent Troop Pay Raise, Cuts Housing Costs," American Forces Press Service, November 26, 2004.

Gilroy, Curtis L., J. Eric Fredland, Roger D. Little, and W. S. Sellman, eds., *Professionals on the Front Line: Two Decades of the All-Volunteer Force,* Washington, D.C.: Brassey's, 1996.

Greenewalt, Crawford H., "I Have Concerns," letter to Thomas Gates, December 31, 1969.

Hardyman, Christine O., ed., *Department of the Army Historical Summary: Fiscal Year 1981*, Washington, D.C.: U.S. Army Center of Military History, 1988.

Helmly, James R., "Readiness of the United States Army Reserve," letter to Chief of Staff United States Army, December 20, 2004.

Hershey, Lewis B., *Selective Service in Peacetime: A Report to the President*, Washington, D.C.: Selective Service System, 1942.

Hoffman, Frank G., "One Decade Later: Debacle in Somalia," *Proceedings: The U.S. Naval Institute*, Vol. 130, No. 1, January 2004, p. 66.

Holtz-Eakin, Douglas, *An Analysis of the U.S. Military's Ability to Sustain an Occupation of Iraq*, Washington, D.C.: Congressional Budget Office, 2003.

Horowitz, Stanley A., and Ayeh Bandeh-Ahmadi, *2004 Defense Economics Conference: Informing the Debate on Military Compensation*, Arlington, Va.: Institute for Defense Analyses, 2004.

Hosek, James R., Jennifer Kavanagh, and Laura Miller, *How Deployments Affect Service Members,* Santa Monica, Calif.: RAND Corporation, MG-432-RC, 2006.

"Impressment," in *The Columbia Electronic Encyclopedia*, New York: Columbia University Press, 2005.

Irondelle, Bastien, "Civil-Military Relations and the End of Conscription in France," *Security Studies*, Vol. 12, No. 3, Spring 2003, pp. 157–187.

Janes, W. Scott, ed., *Department of the Army Historical Summary: Fiscal Years 1990 and 1991*, Washington, D.C.: U.S. Army Center of Military History, 1997.

Janowitz, Morris, "The Logic of National Service," in Sol Tax, ed., *The Draft: A Handbook of Facts and Alternative*, Chicago: University of Chicago Press, 1967, pp. 73–90.

Jeffcott, George F., *United States Army Dental Services in World War II: Demobilization of Dental Corps Personnel*, Washington, D.C.: Office of the Surgeon General, 1955.

Jumper, John P., "The Future Air Force: Remarks by the Chief of Staff of the Air Force at the Air Force Association Air Warfare Symposium," Orlando, Fla., 2003.

Kerr, Bryan, "The 19th Century British Military," Web page, undated, http://athena.english.vt.edu/~jmooney/3044annotationsh-o/military.html (as of May 2006).

Kreidberg, Marvin R., and Merton G. Henry, *History of Military Mobilization in the United States Army, 1775–1945*, Washington, D.C.: U.S. Government Printing Office, 1955.

Lee, Gus C., and Geoffrey Y. Parker, *Ending the Draft: The Story of the All Volunteer Force*, Washington, D.C.: Human Resources Research Organization, 1977.

Lee, Kibeom, Julie J. Carswell, and Natalie J. Allen, "A Meta-Analysis Review of Occupational Commitment: Relations with Person- and Work-Related Variables," *Journal of Applied Psychology*, Vol. 85, No. 5, 2000, pp. 799–811.

Lisbon, Bill, "Commandant: II MEF Slated for Spring in Iraq, Corps Adapts for Terror War," *Marinelink*, July 23, 2004.

Lopez, C. Todd, "Air Expeditionary Forces Concept Is Battle-Proven," Air Force Print News, November 22, 2004.

Lords Spiritual and Temporal and Commons Assembled at Westminster, *English Bill of Rights*, Westminster, England: Parliament, 1689.

Marshall, Burke, *In Pursuit of Equity: Who Serves When Not All Serve? Report of the National Advisory Commission on Selective Service*, Washington, D.C.: National Advisory Commission on Selective Service, 1967.

Molino, John M., *A New Social Compact*, Washington, D.C.: Deputy Assistant Secretary of Defense (Military Community and Family Policy), 2002.

Moore, Albert Burton, *Conscription and Conflict in the Confederacy*, New York: Macmillan, 1924.

Moskos, Charles C., Jr., and Frank R. Wood, "Introduction—The Military: More Than Just a Job?" in Charles C. Moskos and Frank R. Wood, eds., *The Military: More Than Just a Job?* New York: Pergamon-Brassey's International Defense Publishers, 1988, pp. 3–14.

Moskos, Charles C., and Frank R. Wood, eds., *The Military: More Than Just a Job?* New York: Pergamon-Brassey's International Defense Publishers, 1988.

Mueller, John E., "Trends in Popular Support for the Wars in Korea and Vietnam," *American Political Science Review*, Vol. 65, No. 2, June 1971, pp. 358–375.

———, *War, Presidents and Public Opinion*, New York: John Wiley & Sons, 1973.

Nixon, Richard M., *The All-Volunteer Armed Force: A Radio Address by the Republican Presidential Nominee*, Washington, D.C.: Republican National Committee, 1968.

———, "Plan for a Special Commission to Develop Plans to End the Draft," to Melvin R. Laird, January 29, 1969.

Office of the Assistant Secretary of Defense (Public Affairs), "Secretary of Defense Approves Iraq Troop Deployment," news release, Department of Defense, 2004.

Office of Senator Sam Nunn, "Remarks by US Senator Sam Nunn Before the Georgia General Assembly," news release, Washington, D.C., 1973.

Office of the Under Secretary of Defense (Personnel and Readiness), *Report of the Ninth Quadrennial Review of Military Compensation*, Washington, D.C.: Department of Defense, 2002.

Ogloblin, Peter K., *Compensation Elements and Related Manpower Cost Items: Their Purposes and Legislative Backgrounds*, Washington, D.C.: Department of Defense, Office of the Secretary of Defense, 1996.

O'Hanlon, Michael, "Opinion: Nobody Wants a Draft, but What If We Need One?" *Los Angeles Times*, October 13, 2004.

Oi, Walter Y., "Historical Perspectives on the All-Volunteer Force: The Rochester Connection," in Curtis L. Gilroy, J. Eric Fredland, Roger D. Little, and W. S. Sellman, eds., *Professionals on the Front Line: Two Decades of the All-Volunteer Force*, Washington, D.C.: Brassey's, 1996.

O'Sullivan, John, and Alan M. Meckler, eds., *The Draft and Its Enemies: A Documentary History*, Urbana: University of Illinois Press, 1974.

Paine, Thomas, *The American Crisis,* 1776. Available at http://libertyonline.hypermall.com/Paine.Crisis/Crisis-1.html/.

Pike, John, *Fleet Response Plan*, GlobalSecurity.org, last modified April 27, 2005. Available at http://www.globalsecurity.org/military/ops/frp.htm (as of June 2006).

Powers, Rod, "Your Guide to the Military: About Sole Surviving Son or Daughter," 2005. As of May 2006 available at http://usmilitary.about.com/od/deploymentsconflicts/a/solesurviving.htm

Preston, Kenneth O., "Sergeant Major of the Army Preston's Leader's Note Book—November/December 2004," Washington, D.C.: U.S. Army, December 2004.

Putnam's Monthly editorial staff, "The British Army," *Putnam's Monthly Magazine of American Literature, Science and Art*, Vol. 6, No. 32, August 1855, pp. 200–204.

Reeves, Connie L., ed., *Department of the Army Historical Summary: Fiscal Year 1996*, Washington, D.C.: U.S. Army Center of Military History, 1998.

Representatives of the French People, *Declaration of the Rights of Man*, Paris: National Assembly of France, 1789.

Rhodes, Michael L., author interview with Michael Rhodes, Assistant Deputy Commandant for Manpower and Reserve Affairs, November 19, 2004.

Royster, Charles, *A Revolutionary People at War: The Continental Army and American Character, 1775–1783*, Chapel Hill: University of North Carolina Press, 1979.

Rumsfeld, Donald H., *Rebalancing Forces: Review of Reserve Component Contributions to National Defense,* to Secretaries of the Military Departments, Chairman of the Joint Chiefs of Staff and Under Secretaries of Defense, July 9, 2003.

———, *Annual Report to the President and the Congress: Personnel Tempo*, Washington, D.C.: Department of Defense, 2004a.

———, "Foreword," in Barbara A. Bicksler, Curtis L. Gilroy, and John T. Warner, eds., *The All-Volunteer Force: Thirty Years of Service*, Washington, D.C.: Brassey's, 2004b.

———, "Opposition to Re-Institution of the Draft," to Chairman, House Armed Services Committee, October 5, 2004c.

———, *Fiscal 2006 Department of Defense Budget Is Released*, Washington, D.C.: Department of Defense, 2005.

Ryan, Michael C., *Military Readiness, Operations Tempo (OPTEMPO) and Personnel Tempo (PERSTEMPO): Are U.S. Forces Doing Too Much?* Washington, D.C.: Congressional Research Service, Library of Congress, 1998.

Schoomaker, Peter, *Department of Defense Special Briefing on U.S. Army Transformation*, Washington, D.C.: Department of Defense, 2004.

Schwoerer, Lois G., *"No Standing Armies!" The Antiarmy Ideology in Seventeenth-Century England*, Baltimore: Johns Hopkins University Press, 1974.

Senate Appropriations Committee, *Making Supplemental Appropriations to Support Department of Defense Operations in Iraq, Department of Homeland Security, and Related Efforts for the Fiscal Year Ending September 30, 2003, and for Other Purposes*, Washington, D.C.: U.S. Senate, 2003.

Survey & Program Evaluation Division, *November 2003 Status of Forces Survey of Active-Duty Members: Tabulations of Responses,* Washington, D.C.: Defense Manpower Data Center, 2004.

Tax, Sol, ed., *The Draft: A Handbook of Facts and Alternatives*, Chicago: University of Chicago Press, 1967.

Timberg, Robert, *The Nightingale's Song*, New York: Simon & Schuster, 1995.

Timenes, Nick, "Desert Storm Lessons Learned—A Second Opinion," to Assistant Secretary of Defense, May 21, 1991.

Titus, James R. W., and Allan W. Howey, *The Air Expeditionary Force in Perspective*, Montgomery, Ala.: Airpower Research Institute, 1999.

Voeten, Eril, and Paul R. Brewer, *Public Opinion, the War in Iraq, and Presidential Accountability*, Washington, D.C.: George Washington University, 2004.

Washington, George, "To the Committee of Congress with The Army: Head Quarters, January 29, 1778," in John Celement Fitspatrick, ed., *The Writings of George Washington from the Original Manuscript Sources*, Washington, D.C.: U.S. Government Printing Office, 1931–1944, pp. 364–367.

———, "Sentiments on a Peace Establishment," in John O'Sullivan and Alan M. Meckler, eds., *The Draft and Its Enemies: A Documentary History*, Chicago: University of Illinois Press, 1974.

Weiss, Howard M., Shelley M. MacDermid, and Rachelle Strauss, *Retention in the Armed Forces: Past Approaches and New Research Directions*, West Lafayette, Ind.: Military Family Research Institute at Purdue University, 2003.

West, T. D. J., and D. J. Reimer, *A Statement of the Posture of the United States Army: Fiscal Year 1998, Presented to the Committees and Subcommittees of the United State Senate and the House of Representatives, First Session, 105th Congress, February 1997*, Washington, D.C.: Department of the Army, 1997.

Wheeler, Candace A., *Serving the Home Front: An Analysis of Military Family Support from September 11, 2001 Through March 31, 2004*, Washington, D.C.: National Military Family Association, 2004.

White House Press Secretary, *Extension of the Draft and Increase in Military Pay*, Washington, D.C.: The White House, 1971.

Whitely, Gerald, "The British Experience of Peacetime Conscription," *Army Quarterly and Defense Journal*, Vol. 117, No. 3, July 1987.

Whittle, Richard, "Reserve Short on Recruits," *Dallas Morning News*, December 14, 2004.

Williams, Cindy, *Transforming the Rewards for Military Service*, Cambridge Mass.: Massachusetts Institute of Technology, Security Studies Program, 2005.

Wilson, Woodrow, *1915 State of the Union Address*, 1915. Available at http://www.infoplease.com/t/hist/state-of-the-union/127.html.

Wiltse, Charles M., *Medical Department, United States Army: Personnel in World War II*, Washington, D.C.: Office of Medical History, Office of the Surgeon General, 1955.

Winkler, John, *Rebalancing Forces: Easing the Stress on the Guard and Reserves*, Washington, D.C.: Department of Defense, Office of the Deputy Assistant Secretary of Defense for Reserve Affairs, 2004.

Wisener, Terryl L., *Data Book for the All-Volunteer Force—Volume I: The Total Force*, Washington, D.C.: Department of Defense, Office of the Assistant Secretary of Defense (Manpower, Reserve Affairs and Logistics), 1979.

Zinni, Anthony C., "A Commander Reflects," *Proceedings* (*U.S. Naval Institute*), Vol. 126, No. 7, July 2000, pp. 34–37.